D0804198

DNA
USA

Also by Bryan Sykes

Saxons, Vikings, and Celts

Adam's Curse

The Seven Daughters of Eve

DNA USA

A Genetic Portrait of America

Bryan Sykes

LIVERIGHT PUBLISHING CORPORATION

A Division of W. W. Norton & Company

New York • London

Note: Names of some characters from Hollywood films have been used as pseudonyms for DNA respondents throughout the book. These references appear in quotation marks.

Credits: *DNA USA* volunteers thumbnail portraits: Dr. Justin Barrett, Aaron Gray, Jean-Luc Jucker, Dr. Rick Kittles, Dr. Jay Lewis, New England Historic Genealogy Society, Meriwether Schmid, Bryan Sykes; journey images: Bryan Sykes; chapter opener sketches and U.S. location map: Richard Sykes.

For information about permission to reproduce selections from this book, write to Permissions, Liveright Publishing Corporation, a division of W. W. Norton & Company, Inc., 500 Fifth Avenue, New York, NY 10110

For information about special discounts for bulk purchases, please contact W. W. Norton Special Sales at specialsales@wwnorton.com or 800-233-4830

Manufacturing by Courier Westford
Book design by Chris Welch
Production manager: Devon Zahn

Library Of Congress Cataloging-In-Publication Data

Sykes, Bryan.
DNA USA : a genetic portrait of America / Bryan Sykes.
p. cm.
Includes bibliographical references and index.
ISBN 978-0-87140-412-1 (hardcover)
1. Human population genetics—United States—Popular works.
2. Human genetics—Popular works. I. Title.
GN290.U6.S95 2012
559.93'50973—dc23

2011053182

Liveright Publishing Corporation
500 Fifth Avenue, New York, N.Y. 10110
www.wwnorton.com

W. W. Norton & Company Ltd.
Castle House, 75/76 Wells Street, London W1T 3QT

1 2 3 4 5 6 7 8 9 0

For Ulla

Contents

Preface

I am in Duke Humfrey's library, the oldest in Oxford. My desk is made of strong, almost fossilized oak, its surface scratched by the marks of scholars long dead. Before me great books bound in fawn leather blotched with age are arranged on tiers of dark shelving. No reader is allowed to touch them and, save for an occasional dusting, they have rested here undisturbed for centuries. Looking up, my eyes stretch to a ribbed wooden ceiling, its intersections decorated with carved wooden crests: an open book surrounded by three crowns with, underneath, the Latin text *Dominus Illuminatio Mea*—the Lord Is My Light. On the walls hang portraits of archbishops and kings. This is no science library, and there are definitely no books about America. Virtually nothing has changed in Duke Humfrey's library since it was finished in 1488, four years before Columbus set sail across the Atlantic.

When the desk in front of me was first installed, the only voices heard in America belonged to the indigenous people whose ancestors had lived there for thousands of years. When stonemasons carved the joints that still support the roof above me, no Europeans had settled in the New

Radcliffe Camera, Oxford. Part of the Bodleian Library.

World. When carpenters' chisels fashioned the crests of the University adorning the ceiling, no African slaves had been taken in chains across the Atlantic. In the span of only five hundred years, while Duke Humfrey's library has remained as it was, America has become the wealthiest nation on earth and home to more than three hundred million people. Since the library first opened its doors, people have converged on America from all over the world, joining the first Americans and sharing their vast continent.

Hundreds, thousands, of books have been written about the transition from a thinly populated and rural land to the most technologically advanced country in the world. Unlike these myriad accounts, the story I wanted to tell would not be narrated by writers' pens but by a different principal witness, one that is buried deep within our cells and whose voice, until very recently, has not been heard. It is the song of the genes, linking every single American back to his or her origins, wherever in the world those may be. It is also a story told through the experience of the tens of thousands of individuals who have listened to their own ancestral music during the last decade, and what it means to them.

Sitting in the library, contemplating the task ahead, I knew that this was a project on an entirely different scale from any I had undertaken before. The sheer size of the country and the magnitude of the population ruled out any kind of systematic survey. I had to be selective or be overwhelmed. As a geneticist, what I found most appealing about America was that it is the place where the genes from three great continents converge. Like a tidal maelstrom, great currents of DNA slide into one another with powerful force. In the silence of the library, I imagined this struggle as a piece of music, alternating between thunder and serenity like a grand opera or a great symphony.

That is why *DNA USA* came to be written in three movements. In the first I would look back to the original ancestral homelands from which Americans have come. In the second I would explore how these colliding ancestries have been blended in modern America. In the third and final movement I would look deep into the DNA of Americans and see how the atoms of ancient ancestry work together to preserve and alter human destiny.

These matters decided, I closed my notebook, walked down the flights of stone stairs and out into the Bodleian quadrangle bright with May sunshine. What had a few hours earlier seemed impossibly ambitious was now reduced to more manageable proportions. I had a plan. Now all I needed to do was to carry it out.

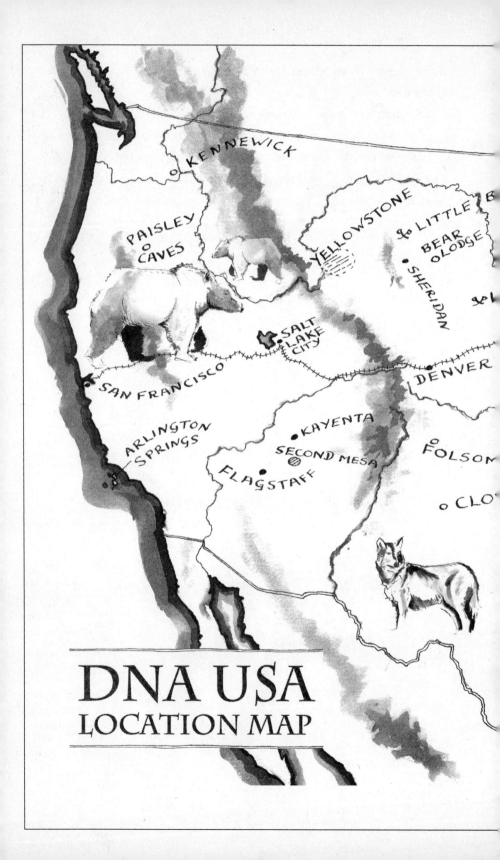

DNA USA
LOCATION MAP

FIRST
MOVEMENT

1

The Point of Clovis

have had a weakness for grand museums ever since I was a boy. I
used to catch the train up to London at least once during each school
holiday and spend the day in South Kensington at either the Natural
History Museum or the Science Museum. Like most boys, my favorite
exhibits were always the biggest: the enormous hissing steam engine or
the giant blue whale, and of course the dinosaurs. As I have grown older
I appreciate these museums more for their embodiment of optimism and
curiosity about the world than for the splendor of their individual exhibits. America has its fair share of grand museums, nowhere more so than
the great showplaces surrounding New York's Central Park or lining the
National Mall in Washington, D.C.

In London none is more celebrated than the British Museum in
Bloomsbury, particularly since its lavish restoration. It houses the main
national archaeological collections, and I spent many happy days there
researching my earlier books. Imagine my delight, therefore, when
one morning I found myself listening to the museum's director, Neil
McGregor, talking about one of the hundred objects from the museum's

Clovis spear point. British Museum, London.

collection that he had chosen to illustrate his radio series on the history of the world. That day he had selected an artifact from the North American Gallery—an object that, as he explained, was evidence that people had arrived in America much earlier than anybody thought. Compared with other displays in the gallery, the magnificent feathered headdresses and the elaborately carved totem poles, this object was at first sight rather modest, but the story it had to tell was of enormous significance for the early history of America. It was a stone spear point thirteen thousand years old, found in Arizona. Dr. McGregor took it out of its case and to the nearby study room, where he described its features in minute detail.

I knew at once that I wanted to do the same and, three months later, after following a trail of connections, I was walking between the Doric columns that frame the museum's famous entrance. Turning right through two glittering galleries, I arrived at a pair of tall wooden doors that led to the back rooms. Once behind those doors, the grand architecture was the same, but the paintwork was chipped and faded. This was the academic heart of the museum, where such decorative luxuries are deemed to be unnecessary. I was met by the archaeologist and curator of the North American collections, Jill Cook, who led the way into the book-lined study room, where a plastic box was waiting for me and, without further ado, the spear point was placed, very carefully, in front of me, cushioned on a sheet of gray foam rubber.

It was about two and a half inches long and a little over an inch wide. Its shape reminded me of a broad fish knife narrowing to a sharp point at one end and with a concave surface at the other. It was a creamy beige, more the color of milk toffee than anything else, and shiny. The first thing that struck me was how very symmetrical it was, and how beautifully made. The workmanship was of an extremely high standard, each face whittled down by a series of blows that had chipped off thin flakes of flint. With the hand lens that had also been brought, I examined the edge of this remarkable stone, but only after I had run my finger along it. It was sharp, but not razor sharp. The edge had been precisely shaped by hundreds of tiny chips, only really visible with the help of the lens. At the broad end there was a shallow groove, about half an inch long, where flint had been removed in a series of strikes that left serried rows of ripples, like a sandy beach after the tide has gone out. Jill explained that this was to help attach the stone point to a wooden shaft without the need for a sinew fastening. By splitting the end of a straight branch, the point could be pushed in along the groove with the ripples of flint acting like a ratchet to stop it from being pulled backward. The attachment was so strong that once the point was in the shaft, Jill did not have the strength to pull it back out.

The spear point I held in my hand had been found in Arizona in 1942, but its particular style derived its generic name from a small town on the border of Texas and New Mexico. That town is Clovis, and the object in my hand was a Clovis point. The site of the town was originally called Riley's Switch, but the coming of the Santa Fe railroad in 1906 created a larger settlement, and to mark its new status, the engineers were asked to suggest a new name. The stationmaster's daughter was enthralled by French early medieval history and came up with the suggestion of naming the new town after one of her heroes, the first Catholic king of France, who had been converted to Christianity in the fifth century AD. Clovis in turn gave its name to the spear points first found in a dry riverbed just south of the town called Blackwater Draw, in 1933. In this way the classic early style of stone tools in America came to have a French pedigree.

The discovery of the Clovis points had a dramatic effect on the prevailing view among the academic community, which at the time was struggling to acknowledge an Ice Age date for the first human settlement in America. The finds at Blackwater Draw were to push the date even farther back into prehistory, but they were already edging in that direction thanks to a chance discovery by a sharp-eyed ranch foreman called George McJunkin a quarter of a century earlier near the village of Folsom, New Mexico, 150 miles north of Clovis.

McJunkin was a forty-eight-year-old African American. Having been born just before the outbreak of the Civil War, he had drifted west after the Union victory that secured his freedom. He was an acute observer of the natural world and took in the abundant signs of life, past and present, in the undulating plains that were his workplace. One day in August 1908 he was on his rounds when he rode into Wild Horse Arroyo, a small canyon still flooded from a torrential storm two days before. The rains had washed away a section of the canyon wall, and McJunkin noticed some large animal bones protruding from the newly exposed cliff. He rode over to take a closer look. They were far too large to be cattle bones and too big even for bison, at least the bison that lives in America in modern times. Though he did not know it at the time, McJunkin had come across the remains of the extinct *Bison antiquus*.

He dug out one of the bones and took it back home. The next day he showed it to his friend, Carl Schwacheim, who he knew was also interested in natural history. Schwacheim, a blacksmith in nearby Raton, tried his best to take this discovery up to a higher level when he got in touch with the director of the Colorado Museum of Natural History. The response was, let us say, less than immediate. It took another eighteen years, until 1926, for anyone from the museum to visit the site at Folsom, and that only came about because the curator, Jesse Figgins, had decided that the museum needed a complete *Bison antiquus* skeleton to put on display. While he was excavating the bison bones, Figgins made the discovery that would shift the debate about the antiquity of the first Americans to a much earlier time. There, within the collapsed rib cage of an enormous bison he found the undeniable evidence that humans had been in at the kill. A flint spear point, its edges knapped with great skill, lay among the bones, witness to a scene played out ten thousand years before.

We can imagine the scenario: The men were ready, lying in wait along the walls of the narrow canyon. They knew from years of experience that the bison would come at this time of year as they moved down from the summer grazing high up in the foothills of the Rocky Mountains to the plains where they would spend the winter. The hunters knew the time was right because the dazzling star Sirius had just begun to rise on the eastern horizon as the sun was setting. Like a diamond in the sky, the Dog Star flashed red, yellow, and ice-hot blue as it edged farther and farther into the heavens. This was the signal for the band to assemble at the killing point, where the canyon narrowed sufficiently to force the beasts into single file. Unlike today, Wild Horse Arroyo was filled with damp vegetation, and a stream ran along the bottom, but it was still a fine place to spring a trap. So important was the bison hunt that the band could not afford to be late and so they waited, relaxing while one of them took up position on the mound half a mile away to warn of the approaching herd. Nowadays we would be bored after a few hours, but boredom was a luxury that never featured in the lives of our ancestors. I say *our* ancestors because this scene, or something very similar, was also being played out in Europe and Asia at the same time. For the members of the band, about twenty strong, were not bored, only patient. The children threw pebbles into the stream that ran along the canyon floor, parting the reeds to discover frogs that hopped back into the cover of the vegetation. Occasionally they would disturb a rattlesnake and, well aware of the danger, taunt it with sticks as it coiled and shook its scaly tail.

Waiting was a skill our ancestors had perfected, but the time was far from wasted. While the children played by the stream, the adults were making sure thay they were ready for the moment to come. The men unfolded the squares of deerskin that held their principal weapon, the glistening flint spear points that would soon be fixed to long sticks of fire-hardened cottonwood. They had been packed away six months earlier in the spring when the bison had reversed their journey on the way to their summer grazing grounds. Now the men took each of their points in turn and ran their fingers along the edges, tapping them expertly with a bone pick to remove a tiny flake here and there and renew the cutting surface. They tested the sharpness of the edge against their thumbnails. If it dug in rather than slide across the surface, the edge was sharp enough. And sharp it had to be to slice through the tough hide of a bison, through a gap in the ribs and into the beating heart of the great beast. Once the men were satisfied, the points were wedged into notches cut into the end of hardened stakes. They were not tied with sinew; any binding would only slow the passage of the weapon through flesh. The journey was only one way, and if the spear was withdrawn for any reason, the point detached and remained where it was.

After three days of waiting the lookout heard something. A low, intermittent rumbling, like far-distant thunder. Peering into the blazing light of the eastern sky he could just make out a faint wisp of smoke rising over the plains. But he knew better. This was neither fire nor thunder but the signal the band had been waiting for so patiently: The bison were on their way. Quickly he ran down to the canyon, spoke, and pointed in the direction of the approaching herd. Immediately the men stopped what they were doing and ran to their positions. Two women climbed to the lookout ready to give immediate reports on the herd's position. The other women ran to the stream and collected the children, moving them to safety away from the base of the canyon that would soon be trampled flat by pounding hooves.

As they waited for the ambush, adrenaline pushed up their heart rates in anticipation. And yet it was not absolutely certain that the bison would choose to come through the canyon. Five years before, after a particularly dry summer, the bison's route to their winter quarters took them three miles to the west, avoiding the ambush altogether. That winter, without the autumn harvest of bison meat dried to last, the band was continually hungry. Three elderly members, well into their forties, did not survive.

For fear that they would be seen and divert the herd, the two women lookouts lay flat and watched the dust cloud as it meandered slowly in their direction. They could not see the animals, but the low, throbbing sound was getting louder. They looked toward the fold in the hills through which the herd usually came. One hour passed, then another. The drumming came and went with the breeze, but still no bison appeared. One of the women climbed down a few feet behind the crest of the hill so she could give a visible signal to the waiting hunters without alerting the herd.

And then they came. Led by three enormous males, the great beasts cantered forward, their black eyes looking from side to side. Across the scorched grass that led to the mouth of the canyon, they came on. The lookout whispered to the messenger words that defined the size and composition of the herd. She then stood up and, using several arm movements, transmitted this intelligence to the hunters below. Her signals indicated that the herd was about forty strong and included several young bison. Having been born in the early summer, these animals would now be about five months old and weigh three hundred pounds.

As they neared the mouth of the canyon, the lead males slowed to a walk, sniffing the air and clearly nervous about entering its narrow confines. The hunters pressed their bodies closer to the rock. A breeze coming down the canyon filled their nostrils with the heavy smell of bison, but more important still, meant that no human scent reached the sensitive noses of their intended prey. They let the big males past, then a few more, until a natural gap appeared in the following animals. Then the trap was sprung. Three young men leaped from their positions on the canyon wall right into the path of the bison, yelling and waving their arms. The forward party accelerated to a gallop and thundered down and out of the canyon. The others shuddered to a halt, fearful of running the gauntlet of screaming humans. They tried to turn and run back up the canyon, but it was too narrow, and their way was blocked by the other animals still pouring forward, unaware for the moment of the melee in front of them.

It was at this point, with the animals uncertain what to do, that the older men leaped down into their midst and thrust their spears into the chests of the panic-stricken beasts: To the low grunts of confusion were added the high-pitched bellows of pain. Two of the speared bison fell where they stood, the flints having found their target of the heart. Others appeared unaffected by the missiles, except for the stabs of pain that

panicked them into trying to push past the bodies of their fallen companions. The men had returned, each armed with a fresh spear, and continued with a second round of slaughter, this time concentrating on the
young animals whose bewilderment was all too plain to see from their
rolling eyes. Three of them fell instantly, the other two reversing out of
the canyon with the spears snapping off as they broke against its hard
walls. The young men, armed with spears, followed them back onto the
flat grassland and ran after them, knowing that eventually they would be
brought down by loss of blood.

That was enough. The spearmen climbed silently back up onto the
canyon walls. The panic subsided, and the twenty or so animals left alive
managed to get past the bodies of the fallen and out of the canyon. It was
never the band's intention to kill the entire herd. They knew very well
that they needed to leave a majority to replenish the herd for the future.
Another of the young men followed a wounded female as she stumbled
down the canyon. It would not be long before she, too, collapsed. All told
the band had killed eight bison: Two adult males were killed instantly;
of the five young, three were killed in the ambush and two others mor-
tally wounded and followed to their deaths a few miles distant; and the
wounded female who fell to her knees soon after reaching the canyon's
mouth. As the sun went down behind the mountains in the west, the
campfires were burning and the air was filled with the mouthwatering
smell of roasting meat. Their faces lit by the flickering light of the fire,
the hunters bade farewell to their quarry as their spirits floated upward
into the sky.

The bones of the slaughtered bison lay where they were. The hunters moved on down to the river valley for the winter. A sudden rain-
storm collapsed the canyon wall onto the bones and there they stayed,
undisturbed for millennia, until another flood washed them out of their
resting place. Though the details of the ambush thousands of years ago
are imagined, there is no invention in recounting what happened next.
Sifting through the bones, Jesse Figgins brushed the congealed yellow
dust from a dark shape, and within five minutes held in his hand the

same flint spear point that had been thrust into the rib cage of the living, snorting beast so many years before.

The effect of finding the Folsom spear point in Wild Horse Arroyo in the summer of 1926 was explosive, but the fuse was slow to burn. At first the find was dismissed by the experts at the Smithsonian Institution in Washington because the excavation had not been performed by an experienced archaeologist. In their opinion this meant that it was not certain that the bison bones and the spear point were in exactly the same layer, and therefore absolutely contemporaneous. To his credit Figgins, who acknowledged his lack of proper training, took notice of the objections and returned to the arroyo. The following year he found what he was looking for, another point embedded in the rib cage of an extinct bison. This time he left it where it was and sent a telegram to the Smithsonian and other leading museums. The Smithsonian immediately dispatched its own archaeologists, who soon arrived to see the evidence for themselves. Naturally eager to discover their own artifacts, within a few weeks they had unearthed an additional seventeen spear points alongside the bones of a total of twenty-three bison in and around the canyon. There was no longer any room for doubt. The first Americans had been there a very long time indeed and were not, as the weight of scientific opinion believed at the time, far more recent arrivals.

Seven years later the time line had been pushed farther back when the excavations at Blackwater Draw, still in New Mexico but almost two hundred miles south of Wild Horse Arroyo, got under way in 1933. The scorching winds that today scour the area had uncovered the gravel bed of an ancient river, long since gone, and among the bison bones the archaeologists discovered an array of Folsom-style spear points. However, beneath the Folsom layer there was another, containing the bones not just of bison but of mammoths too. Here, lying next to the ribs of one of these huge creatures, was another spear point, larger and heavier than its Folsom successor, but equally finely manufactured. This was the Clovis point, just like the one I held in my hand. And since mammoths had disappeared from America by the time the Folsom points were being made, the Clovis point must be from an even earlier time.

In the succeeding decades Clovis points have been found all over North America. There are several examples in the collection that belonged to that inveterate antiquarian Thomas Jefferson, though he had no idea how very old they really were. The widespread distribution of Clovis-style points shows that the hunters, or at the very least their technology, had spread rapidly across the whole continent, and into South America as well. Effective spear points can only be made from stone with the right properties of fracture and hardness, usually flint or chert. Clovis points are often found hundreds of miles from the closest sources of tool stone, and in some parts of the eastern United States more than a thousand miles separate archaeological sites where spear points have been found from the nearest location of the stone from which they were made[1]. All this goes to show that, as in Stone Age Europe, long journeys, maybe along established trade networks, were also commonplace in early America. If this comes as a surprise, it is only because we constantly underestimate the capabilities of our ancestors.

Accurate dating of the bones in which the Clovis points were embedded to about 11,500 years BP (Before Present) unsurprisingly raised the question of whether this date marked the time when the first Americans arrived. Equally unsurprising was the animosity of the exchanges between those archaeologists who believed in "Clovis First" and their opposite numbers, who considered the manufacturers of these elegant weapons to be latecomers, and that America was settled much earlier. Two archeological sites have come to epitomize the evidence in favor of the latter school of thought, the Meadowcroft Rockshelter in Washington County, Pennsylvania, and Monte Verde in Chile. There are several other sites for which claims of pre-Clovis occupation have been made, but it is strongest in these two.

The sandstone shelter at Meadowcroft is located forty feet up on a south-facing slope of the densely wooded valley of Cross Creek, a tributary of the Ohio River. It is forty-five feet wide and equally high, but only twenty feet deep, thus meriting its description as a rock shelter rather than a cave. It was when, as a young archaeology lecturer at the University of Pittsburgh, John Adovasio was looking for a largely untouched site to demonstrate excavation techniques to his classes that he heard about the rock shelter on the nearby Meadowcroft property. Checking on its suitability for his purpose, and before he began any excavations, he found the floor littered with distinctly modern artifacts: beer cans, syringes, and a hash pipe. Clearly Meadowcroft had lost none of its appeal as a temporary shelter or hideout of a sort.

That was in 1973. Adovasio and his colleagues spent the next six seasons working on Meadowcroft, logging an impressive 417 excavation days of twelve to fourteen hours each. Ever since, he has been defending himself and his findings against the school of "Clovis First." Excavating the successive layers of the rock shelter floor, he and his students came across an array of familiar and unsurprising objects down to a level of about ten thousand years, at which point the floor was littered with rocks from a major roof collapse. The feature that made Meadowcroft special was the collection of artifacts that Adovasio found *beneath* the layer of fallen rocks. The first object he and his students came across under the rocks was a spear point. It was about three inches long, and its cutting edges had been repeatedly sharpened by retouching, the process of knocking off small chips of stone with a bone or wood hammer to reveal a new sharp edge. However, it was not of the familiar Clovis design. For one thing, it lacked the fluting that served to anchor the Clovis point so firmly to the split wooden spear shaft. Adovasio named this new spear point the Miller point, after the owner of the Meadowcroft property, Albert Miller, who had encouraged the excavation of the rock shelter. Continuing the excavation into deeper levels, Adovasio found other tools, of presumably older date, made of both bone and stone. The deepest layer contained a two-inch-long stone knife, far more crudely made than the Miller point, but an effective cutting tool nonetheless. This he named the Mungai knife, after a farm a few miles to the east.

Although Adovasio believed from the style of the tools he had found that he was excavating much older layers than the eleven-thousand-year-old Clovis points, he needed an independent method of dating the finds. Fortunately he also came across the remains of campfires that, unlike the stones themselves, could be carbon dated. Briefly, carbon exists in nature in two forms, a stable form with atomic weight 12 (or C^{12}), and a mildly radioactive form with atomic weight of 14 (or C^{14}). All living things, be they bacteria, plants, or humans, contain a lot of carbon. While alive, we and all the rest are literally radioactive, but as soon as we die the unstoppable process of radioactive decay slowly reduces the amount of C^{14} in our remains while the amount of C^{12} stays the same. By comparing the level of these two forms of carbon in the remains, archaeologists can estimate how long ago any organism died. At Meadowcroft as elsewhere, the campfires contain charcoal, the carbonized remains of burned wood cut down at, or shortly before, the time the fire was lit. The oldest carbon date, at the level where Adovasio found the Mungai knife, was sixteen thousand years BP.

No sooner had the Meadowcroft findings been published, in the prestigious British scientific journal *Nature*, than the attacks began.[2] Briefly, there were three separate charges: First, the carbon dates could have been severe overestimates because, so critics argued, flecks of coal might have contaminated the charcoal. As coal, a fossil fuel, is at least 100 million years old and thus completely devoid of radioactive C^{14} which has long since decayed, even a small amount of contamination would have pushed back the carbon dates for the charcoal. Second, and this is a familiar criticism of any excavation, the layers might have been disturbed by burrowing animals or plant roots, thereby burying the stone points at a deeper "older" layer. Third was the climatic argument that sixteen thousand years ago Meadowcroft was only a few miles south of the Laurentide ice sheet that covered most of North America, and thus the area was too cold for human survival. Adovasio answered his critics on all these points. Experts on the analysis of sediments found no trace of coal, the layer of rock from the roof collapse effectively sealed off the lower layers and would have prevented any vertical movement, and, lastly, Adovasio argued that living near the ice sheet was perfectly possible, just as many people live close to glaciers today.

The other archaeological site that seriously challenges the "Clovis First" school lies some five thousand miles distant from Meadowcroft, on the banks of a small creek in Chile. While John Adovasio was excavating in Ohio, another American archaeologist, Tom Dillehay, was teaching in the Universidad Austral de Chile in Valdivia, about five hundred miles south of the capital, Santiago. One day a student brought him the tooth of a mastodon, an extinct smaller relative of the mammoth. Along with the tooth he also brought in a collection of other bones on which Dillehay noticed some cut marks that were consistent with deliberate butchering. Dillehay traveled to the site at Monte Verde, which lies on the banks of the Chinchihuapi Creek about ten miles from the coast of a marine inlet near the town of Maulin. Resting in the sandy terraces on the banks of the creek, he found more animal bones with apparent cut marks as well as stone tools and the always helpful remains of campfires. Imagining these to be no more than ten thousand years old, he was very surprised when the carbon dates from the charcoal and the bones came back at around 13,800 years. Older than Clovis certainly, but still in the same ballpark.

Working their way through earlier levels, Dillehay and his colleagues discovered more stone tools. Most of them were not reworked and looked as though they might have been found and selected as usefully sharp objects. But a few did show clear signs of improvement, in particular a piece of fine-grained basalt about four inches by two and with unmistakable signs of deliberate flaking. This was far less sophisticated than a Clovis point, being worked only on one side and lacking its beautiful symmetry and fluting. Of course, the key question was, as usual, how old was it? Once again it was the ashes from two hearths lying close to where the stones were found that gave the answer, and it was completely stunning. Flecks of charcoal from one hearth came back from the radiocarbon lab at 33,730 years old, and a small pieced of burned wood from the other gave a carbon date of between 33,020 and 40,000 years. In other words, three times as ancient as the Clovis points.

The effect was dramatic, and, just as at Meadowcroft, the critics were ready with a barrage of alternative explanations that undermined what appeared to me to be the careful and understated conclusions of Dillehay's original *Nature* article.[3] As I write, more than twenty years after the paper was first published, the 13,800 years BP dates for Monte Verde have been supported by additional finds, most recently in 2008, and are now widely acknowledged as being correct.[4] However, the significance of the older lower levels at Monte Verde and at Meadowcroft is still extremely controversial. This is despite years of active and often pugnacious defense of their positions by both John Adovasio and Tom Dillehay that has clearly exhausted them both. At various times the debate crossed the boundary from appropriate academic intensity to outright viciousness, with accusations ranging from professional incompetence to downright forgery. Dillehay even went so far as to say that if he had to do it all again, he would not. "It hasn't been worth the agony," he is quoted as saying.[5]

And there, more or less, the matter rests. From time to time other fragments of evidence emerge that suggest that there were people in North America before Clovis. Other archaeological sites, but always lacking actual human remains, surface from time to time. Most recently, in 2008, a site at Paisley Caves in south-central Oregon yielded a few stone tools, butchered animal bones, and human coprolites, the euphemism for fossilized feces.[6] The organic remains, including the coprolites, gave carbon dates of up to 12,300 years BP, slightly older than Clovis but not by much. In another study published in *Science* in 2007, careful redating of material from Clovis and other sites showed that unambiguously identified Clovis artifacts were confined to a very narrow range of only two hundred years or so.[7] The authors were persuaded that this was far too short a time for people to have colonized the vast areas between the scattered sites across America where Clovis points have been found and that therefore America must already have been inhabited and it was the technology that spread. But even that conclusion was fiercely challenged in a strongly worded response to *Science* a few months later, which argued that the new data were completely irrelevant.[8]

The whole debate about the timing and origin of the first Americans has the familiar feel of a stagnant intellectual circus, still balanced between entrenched academic foes who will never agree. This, I have realized over the years, is the natural equilibrium that sets in when a field has reached an impasse and where the rigid stance of personalities and their fiefdoms, rather than evidence, has become the deciding factor in an argument. Although this state of affairs is the antipathy of science as a branch of philosophy, where evidence alone is king, it is surprisingly widespread. When a field stagnates like this, the cycle can be broken only by a completely independent kind of evidence. Which is where genetics comes in.

2

The Nature of the Evidence

lthough scientists have been using genetics to interpret the past for almost a hundred years, it is only in the last two decades that DNA itself has been recruited to the quest. Before that, anthropologists had studied how blood groups and other genetically inherited features varied between people from different parts of the world, and had drawn conclusions about their origins. This was a start, but the tools were very crude and often led to explanations that we now know to be completely mistaken. It is only thanks to the really remarkable technical progress over the past twenty years that scientists have been able to read and interpret the record of the past written in DNA, the language of the genes. In this chapter I will explain how I and other scientists go about doing that. Too many books have been written on DNA for it to warrant a full exposition here, but I do want to spend a little time exploring what makes it such a superb witness of past events—events like the arrival of the first Americans, for example.

In the briefest of terms, DNA is the stuff we get from our parents that instructs our cells to build and run our bodies. It comes, therefore,

A mitochondrion. The powerhouse of the cell.

from an earlier generation. And since our parents received their DNA in just the same way, from their parents, who got it from theirs, it comes down to us from our ancestors, from the past. It is for this very simple reason that the DNA in all of us—the people alive today—can be a witness to events in human history that occurred hundreds, even thousands of years before we were born. How DNA is used and interpreted I shall explain in greater detail during the course of this book. The other remarkable property of DNA is that, under the right circumstances, it is extremely durable and survives for thousands of years in well-preserved fossils. As a witness to past events we have the DNA of living people, which is technically straightforward to analyze but can be hard to interpret, and the DNA of the dead, which is very difficult to work with but considerably easier to interpret.

I still have a vivid memory of the moment when I first saw the cover of the April 18, 1985, issue of *Nature*.[1] On it, in full color, was the fabulously decorated sarcophagus of an Egyptian mummy and underneath it the caption "Mummy DNA cloned." Inside, a relatively unknown Swedish scientist, Svante Paabo, claimed to have extracted and cloned human DNA from one of twenty-three Egyptian mummies from museums in Sweden and Germany. The successful extraction was from the skin of a one-year-old child from the Egyptian Museum in Berlin, which was still in Communist East Germany at the time. What was remarkable about this article was not so much the detail of the result itself, but the sheer gall of the attempt. To geneticists like myself, the thought that DNA could survive longer than a few minutes in the open air was unimaginable. Although not strictly the first extraction of ancient DNA (that was published six months earlier, when DNA was recovered from a museum specimen of the quagga, an extinct relative of the zebra),[2] the effect of the mummy finding was galvanizing. Few people had ever heard of a quagga, but everyone knew about Egyptian mummies especially after the triumphant touring exhibition *The Treasures of Tutankhamun* in the 1970s, and again, as *Tutenkhamun and the World of the Pharoahs*, in 2011.

The visions opened up by the recovery of ancient DNA were completely mesmerizing, and well off the beaten track. In 1985, in my late
thirties, I was enjoying my research in medical genetics at a very exciting
time, truly a golden age, but, surrounded as I was by well-meaning scientists and physicians, I also had a hankering to do something of no conceivable practical value. I can't really explain why that was, but ancient
DNA certainly fitted the bill and, besides, it took me back to my childhood
love of fossils, dinosaurs, and grand museums. Before long I had teamed
up with Oxford archaeologist Robert Hedges to see whether we could get
any DNA from ancient human bone. There was a genuine connection with
my mainstream research at the time, which was investigating the genetic
causes of inherited diseases affecting the skeleton, so it did not raise too
many eyebrows when chunks of bone started to arrive in the lab. Robert and I reasoned that Egyptian mummies, artificially preserved as they
were in the strongly alkaline salt natron, might be a special case. Untreated
bones and teeth were far more abundant in archaeological sites, so if this
new science was to be genuinely useful and widely applicable it was vital
to work out ways of getting DNA from this material. We succeeded after
three of years of trying, and published the results in *Nature* in 1989.[3]

In the intervening two decades between then and now I think it is
fair to say that progress in ancient DNA research has not been entirely
smooth. By far the greatest problem has been contamination of the
material by minute amounts of modern DNA, usually from excavators,
curators, or laboratory staff. DNA does survive for thousands of years
but it gets fragmented in the process. DNA that is damaged in this way is
much harder to recover than youthful, modern DNA—so much so that
even a few molecules of the modern stuff will swamp the damaged original DNA, and you end up with your own or a colleague's DNA sequence.
That's disappointing enough, but not so bad if you recognize it as such.
The real problem comes when you don't recognize it and interpret the
results as being genuinely old.

That happened a great deal in the early days, and I even heard Svante
Paabo once say in a seminar that the DNA he got from the Egyptian

mummy may well have been his own.[4] At first none of us appreciated the extent of the problem. I vividly remember a talk given by a deer biologist who showed a slide of his DNA extraction equipment that he had set up right underneath the head of a white-tailed deer mounted on the wall. No surprise then, that he found deer DNA in the ancient samples he was analyzing. There were some real shockers in the first few years, like dinosaur DNA recovered from a lump of coal and insect DNA from fifty-million-year-old insects trapped in amber. There were certainly many mistaken claims that have never been verified, but they did lead to Michael Crichton's novel *Jurassic Park* and the unforgettable 1993 Steven Spielberg movie of the same name. The almost credible basis of this wonderful piece of science fiction was the recovery of dinosaur DNA from the blood in the stomach of a mosquito trapped in amber that had bitten one of the extinct reptiles. I was asked to the London premiere of the film to comment on it for the evening news on TV. Interviewed outside the film theater in Leicester Square after the screening, I emphasized that even if it were possible to recover genuine dinosaur DNA, putting it all back together and creating a living, breathing *Tyrannosaurus rex* was way beyond current technology.

The next day the lab was invaded by TV news crews who wanted me to repeat the same words for their own bulletins. I eventually got bored with this and arranged for one of my young son's small model dinosaurs to be dragged on a piece of fishing line across the lab bench behind me, just as I put on my most sober expression and announced that I could reassure the public that no cloning of this kind was taking place anywhere in the world. The film crew saw the rubber creature scuttle across the bench and, to their credit, kept it in. The clip was shown that evening, complete with baby dinosaur, without additional comment. Somewhere in the archives the footage rests to this day. But beneath the exuberance, and carelessness, of the first few years, enough results were being independently verified to show that DNA in bones and teeth really did survive for thousands of years. Even if not completely intact, it could be sufficiently undamaged to be extracted and analyzed.

Techniques have improved a lot in the last few years, and early in 2010, the same Svante Paabo, now director of the Max Planck Institute for Evolutionary Anthropology in Leipzig, published the draft genome sequence of a forty-thousand-year-old Neanderthal.[5] Three months earlier a research group from the University of Copenhagen announced the almost complete genome sequence of a four-thousand-year-old Paleo-Eskimo from hair preserved in the Greenland permafrost.[6] As a reflection of the scale of the endeavor needed to do this work, there are fifty-six authors on the Neanderthal paper and fifty-two on the Greenland article. On the original Egyptian mummy paper there was just one author, Svante Paabo himself, while on our 1989 *Nature* article there were only three. Things are now definitely looking up for ancient DNA after a shaky start, but while some technical aspects have definitely improved, others have stayed stubbornly the same.

The press coverage of ancient DNA can always be counted on to be enthusiastic, if not positively melodramatic. The first recovery of DNA was greeted by a memorable headline that gives us a flavor: "US Scientists Clone Dinosaurs to Fight On after Nuclear War,"[7] while the latest announcements from Leipzig once again resurrected the notion that Neanderthals could be cloned.[8]

While complete genomes, which include all of our twenty-three thousand or so genes, are almost routinely sequenced these days, the early work on genetic reconstructions of the past used a much simpler and rather unusual piece of DNA. This is mitochondrial DNA, or mDNA for short. Unlike most of our DNA, which is confined to the cell nucleus, mDNA is found inside small particles, mitochondria, which live in the fluid, called the cytoplasm, that surrounds the cell nucleus but is still contained within the cell membrane. Mitochondria are thus an integral part of every cell, and their job is to convert glucose into energy using oxygen, leading to their sobriquet as "the powerhouses of the cell." Mitochondria were once free-living bacteria that, hundreds of millions of years ago, were kidnapped and put to work by cells. They have lived there ever since, becoming over time completely dependent on their captors. Mitochondria have their own DNA, which, as a legacy of their bacterial origins, is circular rather than linear like our nuclear DNA.

By a great stroke of luck mDNA has several special qualities that are almost perfectly suited for exploring human evolution. For a start, there is lots of it in a cell, far more than its counterpart in the cell nucleus. For every nuclear gene there are hundreds or even thousands more copies in the cell's mitochondria. That made mDNA the natural choice for ancient DNA work, and it was for this reason that we were first introduced to each other. There was always going to be precious little DNA left in an ancient sample, so it made sense to go for mDNA to have the best chance of success. Even these days, when techniques for recovering DNA are far better than they were, there are still plenty of ancient samples where mDNA is the only DNA you can get out. I always check that I am able to extract mDNA from an ancient sample before embarking on recovering other genes.

While its cellular abundance was reason enough for me to choose mDNA for ancient work, it also comes with two other great advantages. The first of these is that mDNA changes far faster over time than the cell's other DNA in the nucleus. However, the fast rate of mutational change in mDNA is only relative and, to most of us, appears mind-numbingly sluggish. On average mDNA changes at the rate of one mutation every twenty thousand years. Though that is undeniably slow by everyday standards, it is still twenty times faster than the mutation rate of nuclear DNA. Very conveniently, as we shall see, the mDNA mutation rate fits in very well with the timescale of human evolution. The second great advantage of mDNA is that, unlike most nuclear DNA, it is inherited down only one line of ancestors. While human eggs are crammed with a hundred thousand mitochondria, sperm have hardly any, and none at all that survive past fertilization. This means that the mitochondria in a newborn baby have come exclusively from his or her mother. This fact makes the inheritance pattern of mDNA simplicity itself. Every man, woman, and child's mDNA is inherited from his or her mother, who gets it from her mother, who inherits it from her mother, and so on back in time. The direct matrilineal line stretches back hundreds, thousands, tens of thousands of years. All that time the mDNA stays exactly the same, save for any changes introduced by mutation once in every twenty thousand years or so. There is another tremendous advantage of mDNA when it comes to looking at ancestral origins, and especially those of indigenous people, but we will look into that a little later. .

Mitochondrial DNA, like its counterpart in the nucleus, is made up of a series of four slightly different chemicals, called nucleotide bases, strung together in a particular sequence. Four different-colored beads on a string is a useful and workable metaphor here. It is the order of these colored beads that is all-important. However, unlike necklaces that might stretch to a couple of hundred beads, DNA is immensely long. Even mitochondrial DNA is extensive by familiar standards. Its circle of DNA contains more than sixteen thousand of these metaphorical beads, but this is minuscule compared to the entire human genome, which is made up of three thousand million of them. I will explain more about how these bases direct the intricate working of our cells later, but for now the important detail is their precise sequence, which we can read using a variety of techniques. For mDNA we are concerned only with the sequence of its sixteen thousand or so bases and, most important, the fact that this sequence can and does change through the process of mutation. In fact we can narrow down our interest to a stretch of four hundred bases that change, or mutate, more rapidly than the rest because mutations in this segment, called the control region, have no biological consequences. Changes outside the control region can interfere with the efficient working of the mitochondria in such a way that the mutated mDNA is eliminated and not passed on to future generations.

If two people have the same mDNA sequence, then they must have inherited it from a shared maternal ancestor. It is this simple concept that underlies the way mDNA is used to reconstruct the past. Even when the two people concerned live in completely different parts of the world, if their mDNA sequences match, then there must be an ancestral connection between them, even if they are not aware of it. Imagine your own line of maternal ancestors going back through your mother, her mother, and so on back into the past. Now imagine someone else who has a matching mDNA sequence. If you were able to trace his or her maternal ancestral line and your own far enough back in time, they would meet in one woman, your shared maternal ancestor. There is no need for paper records to confirm this ancestral connection. It is there in your DNA.

In this scenario, where two people have matching mDNA sequences, their next question is going to be this: How long ago did the woman, from whom we are both descended through the maternal line, actually live? This question introduces another key element, that of time. While we might occasionally be able to give a precise answer, through establishing a rigorous genealogical connection back to a common ancestor, most of the time this will be impossible. Nevertheless we can arrive at some sort of answer by using the fact that, although mDNA is very stable, it does change slowly over time through the process of mutation.

In the control region these mutations happen when mDNA is copied as cells and their mitochondria divide, which they are doing all the time. Returning to the bead necklace analogy, just occasionally the copying is inexact and a different-colored bead is substituted in the copy. This is a mutation, and as far as we know it is an entirely random event. The only mutations that concern us here are those that occur in the female germ line—that is, in the cells which become eggs. They are the only ones that get passed on to the next generation. All the other mDNA mutations that happen in our other body cells, like muscle and blood, are of no consequence as they will not be passed on and will disappear when we die. In men even germ-line mDNA mutations are irrelevant, since the few that get into sperm cells are destroyed in the fertilized egg and never make it to the next generation.

From a variety of measures, we have a pretty good idea of the mDNA mutation rate, and it is about one every twenty thousand years. If the average generation time for women over the course of human evolution was twenty years, which is not unreasonable, then this is the equivalent of roughly a thousand generations. In other words, in one of every thousand female births, the mDNA of the child will differ by one mutation from the mDNA of her mother. Notice that I haven't counted the changes between mothers and sons because, although they will also occur, they don't matter because the son's mDNA is not going anywhere.

To illustrate how we can use this for genetic dating, let us imagine two people who have mDNA sequences that do not precisely match but differ by one mutation. In fact we can use an actual example. The mDNA of the last Russian czar, Nicholas II, which was recovered from his remains after they were excavated in 1991, differs from mine at only one position. Our shared maternal ancestor must therefore have lived sufficiently long ago for one mutation to have accumulated in one of the two lines of descent leading from that shared ancestor to the czar and myself. Using the average mDNA mutation rate, we can assume that the czar and I are separated by a total of a thousand generations from our common maternal ancestor. That is five hundred generations along one line and five hundred along the other. So, by this reckoning, the woman from whom we are both maternally descended lived five hundred generations ago. If we take twenty years as the generation time, then she lived ten thousand years ago. I have to emphasize here that this is an *average* figure. Our common maternal ancestor may have lived more recently, or longer ago than a thousand generations. This is all down to the random nature of mutation and makes genetic dating of this sort very approximate for individual cases. In this respect it is much less exact than the carbon dating that we covered earlier. But it can be very useful when large numbers of people are compared. And that is how mDNA first came to be used to explore the way in which humans have moved around the planet over the past quarter of a million years.

Regardless of where they were working in the world, as soon as scientists began to look at their results from mDNA analysis they noticed that, far from being uniformly spread, individual DNA sequences tended to fall into distinct clusters. Within the clusters the sequences were clearly related to one another. To give you an example, one of the clusters found among Native Americans, as we shall see shortly, features a mutation at position 16,111. The number 16,111 refers to its place in the sequence around the mDNA circle, starting at an agreed point. The mDNA control region that concerns us here extends between positions 16,000 and 16,400, thus I will drop the 16,000 prefix from now on so that 16,111 becomes 111, and so forth. The mutations are changes relative to an internationally agreed reference sequence, which is that of the very first mDNA to be sequenced in 1981. The DNA base at 111 in the Native Americans is different from the base at position 111 in the reference sequence. In fact no Native American mDNA so far sequenced has only the mutation at 111, but many people have 111 plus one or more other changes. Likewise in Europe, many people share a mutation at position 224 in their mDNA when compared to the reference sequence. In Polynesia almost everyone has a mutation at position 247, on so on.

In the early years of my research with mDNA, I got to know these numbers very well indeed, and it was an exciting time trying to make sense of the different genetic clusters that they defined. In Europe, and among Americans with European roots on their mothers' side, there are seven such clusters and I eventually realized that this had to mean that each cluster has just one matrilineal ancestor. From this it followed by an irrefutable logic that there were only seven women, seven "clan mothers," from whom almost all native Europeans are maternally descended. In time these women became the heroines of my first book, *The Seven Daughters of Eve.*

3

The First Americans

Until the invention of agriculture, which over the last ten thousand years has happened at various times in different parts of the world, our ancestors were on the move most of the time. They needed to be, as their main food supply derived from the herds of game that migrated with the seasons. Bison, mammoth, and wild horse were all on the menu, and on the move, and our ancestors had no choice but to follow. They scavenged the carcasses from the kills of lion and hyena, or killed their own by ambush or by stampeding the herds over cliffs. If there were encampments, they were transient and seasonal. Only in a very few parts of the world were there anything resembling permanent settlements, and one of these was the Pacific Northwest of America. The reason for this luxury of permanence, from which all sorts of benefits flowed, is summed up in one word: salmon.

Then as now, the rivers of the western seaboard of America from Oregon to Alaska were conveyor belts of fish heading upstream to spawn. I once saw this spectacle myself on the Columbia River, where a fish ladder had been built to bypass a hydroelectric dam. Through the glass of the

Leaping Coho salmon.

viewing gallery, where lampreys held on with their sucker mouths and edged up against the current, the silver shapes of Coho salmon appeared, paused for a few seconds, then continued their relentless journey to the spawning grounds with a powerful flick of the tail. These were a good two and a half feet long, weighing maybe seven or eight pounds. Then, suddenly, an enormous fish appeared, well over thirty pounds, I would guess. It stayed for a second or two and then it was gone, leaving me with the feeling that I had only imagined it. Then another came and went. And another. This was the King salmon, the monarch of the river.

Such wealth as this sustained the Indians of the Pacific Northwest for at least four thousand years before they first encountered Europeans. The stability of the food supply, with dried fish filling the gap between annual spawning runs, made for the exuberant flowering of a culture for which the region is still known today. Relieved of the burden of being always on the move and having to carry everything from one camp to the other, their dwelling houses and their ceremonial totem poles became larger and more elaborate.

It was one of these tribes, the Nuu-Chah-Nulth, which became the first to come under the scrutiny of DNA, and in so doing became the tribe against which all others would come to be measured. The Nuu-Chah-Nulth—the name means "all along the mountains"—live nowadays on the western coast of Vancouver Island and on the Olympic Peninsula of Washington State, a hundred miles south of Seattle. There are fifteen surviving groups within the Nuu-Chah-Nulth, far fewer than before European contact and its dreadful legacy of smallpox that so reduced the population. The first European to reach the Nuu-Chah-Nulth was the British navigator and explorer James Cook in 1778. He was on his way to try to find the Northwest Passage rumored to connect the Pacific and Atlantic oceans. Cook sailed on through the Bering Strait, but his way was blocked by sea ice. Ironically, had he tried the same voyage today, he might have got through thanks to the melting effects of the current phase of global warming.

Today the Nuu-Chah-Nulth number around nine thousand. As well as being experienced fishermen, they have also in the past been skill-ful whalers. Along with other nearby tribes, they are enthusiastic par-ticipants in the potlatch ceremony, where hosts distribute lavish gifts in surroundings of celebration, music, and dance. Although potlatch was banned by the Canadian and U.S. governments in the late nineteenth century, the legislation was eventually repealed in 1951, and the cer-emony soon regained its earlier popularity. Between 1984 and 1986 a large proportion of the Nuu-Chah-Nulth from Vancouver Island were recruited by scientists from the University of Utah as part a survey of blood groups and other biological markers. After the project had been completed, the blood serum was frozen, and it was this stored mate-rial that was used for the very first analysis of Native American mDNA sequences. It was also a minor technical triumph in that retrieving DNA from stored serum, which does not contain blood cells, had not been tried before. True, mitochondrial DNA had been analyzed five years earlier from a group of Pima Indians living along the Gila River valley near Phoenix, Arizona, but the data were confined to only a handful of genetic markers rather than DNA sequences themselves, where far more information resides.

Sixty-three individuals were chosen who were known, from the extensive genealogical records collected by earlier researchers, not to be maternally related. Had they been, their mDNA would have been auto-matically identical and would have skewed the interpretation. What the Nuu-Chah-Nulth mDNA revealed was that there was far more genetic variability within the tribe than had been suspected. There were twenty-eight different sequences among the sixty-three Nuu-Chah-Nulth vol-unteers whose mDNA had been analyzed. At one extreme the most frequent of these was shared by nine individuals, while at the other, thirteen people had mDNA sequences found in nobody else in the group. Nonetheless there was a pattern. The Nuu-Chah-Nulth fell into four clusters within each of which their sequences were related to one another. The biggest of these clusters was made up of twenty-eight indi-viduals, all of whom shared mutations at position 111 and 319, variations that were absent from the rest of the group. So they had to be related to each other, and they must have inherited these particular mutations from a common maternal ancestor. There were plenty of other mutations among the people in the cluster, but most of them were confined to a few individuals.

The three other mDNA clusters had similarly shared sequences, and some differences in detail. In one cluster everyone shared a mutation at 278, while the mDNA from members of a third cluster all shared a variant at base number 325. All three of these clusters were united by the crucial mutation at 223, of which more later. The fourth cluster, however, did not have the variant at 223, but instead shared mutations at 189 and 217.

When I read this paper soon after it was published in 1991,[1] my eye caught this last detail. I had only recently returned from my first visit to Polynesia and was working through the analysis of the twenty Cook Islanders whose DNA I had brought back with me to Oxford. Nineteen of the twenty also shared these two mutations. This had to mean they were in some way maternally related to the Nuu-Chah-Nulth, but whether it meant that these Polynesians were descended from Pacific Northwest Indians, or the other way around, took some time to work out. The connection was confirmed a little later when the DNA from both the Cook Islanders and the Nuu-Chah-Nulth was shown also to lack a small section of DNA, nine bases long, which had been deleted from the mitochondrial circle. This deletion was extremely rare elsewhere, and combined with the sequence match, could lead to only one conclusion: They were definitely related.

The message from the Nuu-Chah-Nulth paper was powerfully clear. This group of American Indians from the Pacific Northwest contained not one but four separate clusters of mDNA, each one of which contained masses of individual differences. This was not the DNA picture from a single recent migration of just a handful of people, as had been suggested. The sheer variety of sequences showed that the ancestors of the Nuu-Chah-Nulth were both numerous and ancient. But how old?

Fortunately that was a question that could be answered, although exactly how to go about it took a few years to perfect. And the next question was this: If the ancestors of the Nuu-Chah-Nulth had migrated from somewhere else, where was the source? And what about Native Americans from other parts of the United States, and indeed from South and Central America as well? The rush was on to find out.

Over the next few years scientists pursued the genetic origins of Native Americans with innocent enthusiasm, rarely if ever pausing to wonder what their subjects felt about it. But these were the golden years before such considerations became sophisticated and serious. The clouds of concern lay well beyond the horizon at the time. The work was technically straightforward, and as long as you could get your hands on some DNA, the rewards were both interesting and substantial. Peer-reviewed papers, the medium through which all modern scientists have come to be judged, flowed from the labs with regular ease. The extremely well-regarded *American Journal of Human Genetics* became almost the trade paper of the new breed of genetic anthropologists. Throughout the 1990s, papers on the genetic origins of Native Americans, along with other indigenous peoples from all over the world, were published in almost every issue—at least that's how it felt at the time. I enjoyed publishing my own research papers on Polynesia and Europe in the *American Journal of Human Genetics* because, unlike *Nature* and *Science*, where articles were pared down to an absolute minimum, the *American Journal* (as it was known to all of us) gave you the space to present the detail of what you were doing and to discuss the conclusions at reasonable length.

As far as the research on Native Americans went, before long the kind of study that had been done on the Nuu-Chah-Nulth had been replicated in populations throughout North and South America. From the Inuit of Alaska in the North to the Mapuche of Chile, dozens of papers containing hundreds of individual mDNA results were published, many of them in the *American Journal*. I remember eagerly awaiting each new issue, between the familiar green-and-gold covers, as the librarian put out the copies on display. What treasures did this month's edition contain? There was almost always something.

It was very soon clear that a pattern was emerging from all the effort being put into Native American DNA. First of all, the four clusters seen in the Nuu-Chah-Nulth were also found in virtually all the other studies, though their proportions varied considerably depending on where the DNA samples were from. By then a new vocabulary was emerging to identify these mitochondrial clusters and, for the simple reason that Native Americans were the first populations to be thoroughly studied using the new genetic tools, they were allotted the first four letters of the alphabet: A, B, C, and D. To emphasize that they were real women, and not just theoretical constructs, I gave them each names that, following the convention of *The Seven Daughters of Eve*, began with the initial of the cluster's scientific notation. The clan mother of cluster A became Aiyana, Chochmingwu was the name I gave to the founder of cluster C, while Djigonese was the clan mother of cluster D. For reasons I will explain in a moment, the founder mother of clan B broke with this convention, and I called her Ina.

There was reason to believe that three of these clusters were related to one another because they all possessed the important mutation at position 223. The exception was cluster B. The same logic of a shared maternal ancestry applies to the four American clusters just as it does to the seven European clans. Each cluster has just one woman as its matrilineal ancestor. In Europe I was able to estimate how long ago these seven clan mothers lived by extending the reasoning we saw earlier when comparing two individuals—that is, by averaging the mutations among all individuals within each cluster and multiplying this figure by the mDNA mutation rate. The result of this calculation showed that the seven native European clan mothers lived at different times between ten and forty-five thousand years ago. The same calculation applied to the Native American clusters gave clan ages of between just over thirty thousand years for cluster D and just under ten thousand for cluster C.

A direct extrapolation from these genetic cluster dates to the timing of the first settlement of America appeared to support the idea that members of cluster D had arrived well before the time of the Clovis period around 11,500 years ago. Taken at face value, there was even a possibility from the genetic results that they had arrived as far back as the early and controversial lower levels at Monte Verde, which returned carbon dates of more than 30,000 years. However, things are not quite that simple. The additional consideration is that the only mutations that are relevant are those that happened *after* the mDNA got to America. Any mutations that occurred in matrilineal ancestors before they got to America might tell you something about what was happening in the ancestral homeland, but not in America itself.

To illustrate the point, imagine trying a similar analysis on the Europeans who arrived in America within the last five hundred years. All seven European ancestral clans are represented, and there are lots of individual differences within the clans. But to apply the same method of averaging the mutations within each clan and multiplying them by the mutation rate would give ages for the clans that were pretty much the same as they were in Europe. It would be completely misleading to use these figures to conclude that Europeans arrived in America between ten and forty-five thousand years ago.

After several years of collecting and analyzing Native American mitochondrial DNA, scientists eventually developed a way of distinguishing between ancestral founder sequences and derived sequences brought about by mutations that occurred in America after settlement. Only the derived sequences—mutations that had happened *in situ*—could be properly counted toward arrival time estimates. If a sequence could be found both among Native Americans and in people living today in an ancestral homeland, then it could be assumed that this was one of the founder sequences that should be discounted from the age estimates. If, on the other hand, a sequence had clearly evolved by mutation from a founder sequence, and was found in America but not in the ancestral source population, then it should be included in the age estimate. Founder analysis, as this method became known, is a tricky business that, for a start, requires that you know where the appropriate ancestral regions are.

Ever since the Spanish scholar José de Acosta first proposed an Asian origin for Native Americans in 1596, historians and anthropologists have thought of northeast Asia as the most likely source of the first American settlers. The popularity of this hypothesis only increased when Vitus Bering showed how close the two landmasses really were when, in 1728, he sailed from Siberia to Alaska across the strait that now bears his name. Although there were other theories, as we shall see, there were a number of reasons for making a connection between Native Americans and northeast Asians, and these include some similarities in facial features and tooth shape. The blood-group frequencies of Native Americans and northeast Asians are also similar, so it came as no great surprise when collaborations between Russian and American scientists got under way as Cold War inhibitions thawed, that there were general similarities in the mDNA clusters found among the two groups. The genetic results from both continents were gathered together and put through a founder analysis, individually scrutinizing each Siberian and Native American sequence to identify, as far as possible, when an American mDNA had evolved in America rather than in Asia. This process had the effect of setting the genetic clock to zero so that the age since settlement could be calculated from mutations that had accumulated *in situ* rather than in the source population—which by its very nature must be older. As my research team[2] and I discovered when doing something similar in Europe, founder analysis is a very laborious process. To do it properly a large amount of data are needed from the source population, far more than from the derived, so as to reduce the chances of missing founder sequences.

Among Native Americans, calibrating the genetic arrival dates by founder analysis had the effect of making them more recent than the initial, uncalibrated estimates. The first attempt, in 1996, narrowed the range of genetic settlement dates in North America from 10,000 to 30,000 years to between 18,000 and 25,000 years.[3] The most recent set of calibrated estimates was published in 2009.[4] By then it had become almost routine to sequence the entire mitochondrial genome of more than sixteen thousand DNA bases rather than the short four-hundred-base control region. This extra sequencing certainly showed all possible mutations, but the richness of variation retained in the control region meant that the gain was not proportionate. The increase in mutation data and in the number of individuals involved ironed out the more extreme age estimates from the first founder analysis published thirteen years previously, and gave dates for the clusters ranging from 15,800 to 19,600 years (details are in the appendix). I think we will have to settle for that. In my view there is little scope for improving the accuracy of these estimates with further work.

Insofar as genetic and carbon dates are remotely comparable, it seems to me that these calibrated mDNA dates suggest that the first Americans did arrive before Clovis. The same studies that identified founder sequences also established beyond reasonable doubt that Siberia was the original entry point for the majority of ancestors of today's Native Americans. However the research also uncovered a significant surprise. While a Siberian origin for clusters A, C, and D is clear and the founder sequences identified, the same is not true for the fourth cluster, B.

Cluster B is my favorite because it is predominant in Polynesia, where I began my work with mitochondrial DNA twenty years ago. I don't have to look up the main Polynesian control region sequence—it is etched on my memory: 189 217 247 261. Believe it or not, even now, when I come across this sequence during my current research projects, I am back on the white sand beaches of Rarotonga, in the Cook Islands, gazing out to sea across the surf line as the Pacific Ocean breaks on the reef. Three thousand years ago the very first Polynesians arrived here from the islands of Indonesia across hundreds of miles of open ocean. It was, I believe, the greatest feat of maritime exploration in the history of our species. I must be careful not to go on about this too much, and anyway I have already written extensively about my time in Polynesia in *The Seven*

Daughters of Eve. Suffice it to say that the experience completely altered my view of the accomplishments of our ancestors. I no longer saw them as the savage equivalent of ourselves, but as fearless and resourceful pioneers who survived the many trials of raw nature that would defeat almost all of us living today.

Tracking the origins of the Polynesians to Asia, and not to America as the Norwegian anthropologist Thor Heyerdahl famously argued, was fairly straightforward. I could see that the founder population was from Taiwan, where the sequences 189 217 and 189 217 261 were found among the aboriginal Taiwanese. You will forgive me, I hope, for breaking with convention and naming the clan mother of cluster B after the mythical Polynesian princess Ina, who was carried to her lover's floating island by a shark and whose image adorns the Cook Islands banknotes.

I scanned the mitochondrial sequences of Native Americans as they were published, eager to see if my heroines had managed to cross the entire Pacific Ocean and had taken part in the colonization of America, but I never did find the Polynesian motif. It was too far, and three thousand years ago was too late in the day, for the Polynesians to make any serious impact. We do know that they reached South America and returned with the sweet potato, a decidedly Andean crop, and also that large numbers of Polynesians were enslaved in the nineteenth century and taken by force to Peru.[5] But they did not play a significant part in the settlement of America, with the notable exception of Hawaii, where the unmistakable Polynesian sequence is still to be found among the diminishing number of native Hawaiians.

Although the magical Polynesian "motif" is not found in America, the ancestral sequences certainly are; 189 217 and even 189 217 261 are right there at the center of the cluster B sequences of Native Americans. But how did these sequences arrive in America? Unlike A, C, and D which are widespread among today's Siberians, cluster B is virtually unknown there. That itself is curious, if the only way of getting to America was across Beringia and the land bridge from Siberia to Alaska. Neither is cluster B found more than very sporadically among modern Alaskans

and is only encountered in numbers from Vancouver Island south. However, in Central and South America it is the predominant cluster where,
if anything, the cluster's genetic dates are slightly earlier than they are
in the North. To me and to other scholars, notably another Polynesian
enthusiast, Rebecca Cann from the University of Hawaii, this geographical distribution smacks of a distinctly nautical arrival and suggests that
cluster B arrived in America not by land but by sea.[6] The calibrated
genetic date for cluster B in America is 18,700 years, on a par with the
three other Asian clusters, which almost certainly crossed from Siberia.
But what were conditions were like at the time?

America is separated from the rest of the world by two vast oceans, the
Pacific and the Atlantic, with only fingers of land in Siberia and Alaska
reaching out toward each other, like God and Adam in Michelangelo's
famous mural on the ceiling of the Sistine Chapel in the Vatican. From
time to time over the last quarter of a million years these fingers have
touched and let go again as the sea around them fell and rose with the
cycle of the ice ages. If humans needed dry land to reach America, then
they would have to wait for the ice caps of the Arctic to suck enough
water from the ocean to expose the causeway.

   According to climate records stored in bubbles of air trapped in ice
deep below the surface, the temperature was only low enough for long
enough to expose the land for short intervals during the last two hundred
thousand years. One was during the last Great Ice Age, which reached
its peak about eighteen thousand years ago. This enduring spell of freez-
ing weather lasted for nearly ten thousand years. Glaciers grew from the
mountain ranges of the North, connecting to become vast ice domes
covering much of North America and Europe. The ancestors of modern
Europeans were forced southward to Spain and Mediterranean France,
before the gradual thaw drew the herds of game on which they depended
north again. Earth's orbit around the sun wobbled again about thirteen
thousand years ago, and the conditions of the Ice Age returned, once

more allowing Siberia and Alaska to join hands. But this was a comparatively brief spell and was followed by a rapid warming that announced the stable conditions we have known for the last ten thousand years. If the ancestors of the first Americans arrived only by land then, according to this research, there were just two opportunities to do so, between 23,000 and 17,000 years ago during the Great Ice Age, and later during the mini-ice age, called the Younger Dryas, around thirteen thousand years ago.

During these cold phases the sea was three hundred feet or more below its present level. The tips of Siberia and Alaska were connected by a broad expanse of land up to a thousand miles across, and most people who have pondered the origin of the First Americans have tied their arrival to one or another of these periods when the land connection between Asia and America was intact and open for traffic. However, I do not see any convincing reason for this restriction. Even today, when the causeway is flooded by the sea, the crossing from one side of the Bering Strait to the other is only just over fifty miles. Not only that, the gap between Siberia and Alaska is not all water. The Diomede Islands, one belonging to Russia, the other to the United States, are conveniently situated around the halfway mark. Nowadays caribou and arctic fox routinely swim between the mainland and the islands, and it would not have been dangerous for humans to set out for the Diomedes from Siberia by boat. Once there, Alaska would have been plainly visible.

Our ancestors certainly knew about boats by the time the first humans reached the Arctic shores of Siberia. They had reached Australia across the open sea at least fifty thousand years ago. We do not know for sure whether Siberians had boats or not, and there is scant hope of ever finding any direct evidence: Boats were constructed from biodegradable materials like bone, skin, and wood that have long since turned to dust. It is conceivable that a frozen boat, deliberately buried, might one day be discovered, dating to a time when we know that the Bering Strait was open water, but it isn't likely. Likewise we have no direct proof that the first Australians had boats, only the powerful circumstantial logic that, even when sea levels were at their lowest, any route from Asia to Australia involved a sea crossing of at least thirty miles. And, of course, in the Arctic the sea can freeze over, creating a temporarily solid connection. Even now, in the severest winters, sea ice forms right across the strait and is thick enough to support the weight of a human and a dogsled. Whether on a moonlit ice crossing under the flickering phosphorescence of the northern lights, or by boat in the endless sun of the Arctic summer, the Bering Strait can never have been a serious obstacle to human movement. I see no need at all to restrict the arrival of the first Americans to those periods of time when we know the connection to Siberia was dry land. Our ancestors were far too adventurous to let that stop them.

The curious geographical distribution of cluster B, absent from higher latitudes but abundant in Taiwan to the west and in Central and South America to the east, does raise the possibility that the ancestors of these Native Americans arrived by boat. The east–west flow of wind and currents across the Pacific makes it very difficult to navigate directly from Asia to America across the open sea. Even the oceangoing galleons of Spanish navigators during the fifteenth and sixteenth centuries were forced to travel along the great circle route to return from Indonesia and the Philippines to their bases in Central America. That meant following the Pacific coast north to China, Japan, and Kamchatka, and then crossing the Bering Strait before turning south and skirting Alaska, Canada, and California before reaching their home ports in Mexico. I find it very reasonable that the same route might have been taken by the ancestors of Native Americans in cluster B who set out from the Pacific coast of South China and Taiwan. Even though the waters were cold, fish and sea mammals were abundant just as they are now. Perhaps the cluster is absent from Alaska and the icebound northern Pacific coasts simply because they did not stop, preferring to follow the sun south toward the more inviting climates of Central and South America. This preference might explain the astonishingly rapid movement of the first Americans from north to south, judging by the early archaeological dates from coastal sites in Chile.

Even though, just as in Australia, we have no preserved remains of boats, there is hard evidence of marine travel. At Arlington Springs, on Santa Rosa Island off the California coast near Santa Barbara, human remains carbon dated to 11,000 years BP, but with no distinct stone tools, could only have arrived by sea.[7] Even at Monte Verde, in southern Chile, microscopic examination of the soil around the hearths, dated to 14,600 years, found tiny fragments of edible and medicinal seaweed, showing that the inhabitants were very familiar with the abundance of coastal resources.[8] The same was true for our European ancestors living at the same time and a little later, who preferred the coasts to the interior.

My own summary, based on an amalgamation of the archaeological and genetic results, is that the first Americans arrived by both land and by sea. The mammoth hunters from Siberia, mainly from mitochondrial clusters A, C, and D, arrived in Alaska either across the causeway of dry land when the sea levels dropped, or covered the short distance by boat or across the ice at other times, probably using the Diomede Islands as a convenient stepping stone. Another group, from the descendants of Ina in mitochondrial cluster B, traveled by boat from much farther south. They did not stop to settle in Siberia or Alaska but carried on to the more moderate climate of Central and South America. There is no need, in this reconstruction, to confine the arrivals to different "waves" of settlement; and, as elsewhere in the world, there would have been movement back and forth. Nonetheless, the genetic echoes of the epic Arctic journeys of their ancestors are still there in the cells of today's Native Americans.

Coastal travel down the Pacific coast of America was comparatively easy for the descendants of Ina, but for the Siberians going south by land, the way was blocked by two vast ice sheets. As the orbit and inclination of the earth around the sun shifted slightly around twenty-five thousand years ago, the cycle of ice ages that have been a feature of the earth's history for the last two million years once more turned the thermostat down. At first, snow that used to melt every summer began to resist the sun and lay all year round. As the temperature continued to drop, these patches of permanent snow grew into small glaciers that in turn coalesced into ice fields. Slowly these ice fields grew in thickness and extent until, by twenty thousand years ago, an enormous dome of ice, the Laurentide, covered most of Canada and the United States north of a line joining New Jersey, Cincinnati, and St. Louis, Missouri. Farther to the west, the mountain glaciers of the Rockies and the Cascades grew and fused into a second ice cap, the Cordilleran, that stretched as far south as Seattle. Only the strip of land between this and the larger Laurentide sheet covering the Midwest was free from permanent ice cover, at least some of the time. Ironically, Alaska, outside the McKinley range, was spared this crushing load as it was just too dry for the snow and ice to build up. Overland access to the interior would have been possible through the ice-free corridor, though exactly when and for how long is still controversial. The other way south was along the coast, where a thin sliver of land was kept free of ice by the warming effect of the oceans. Even where glaciers blocked the coastal route, a short diversion out to sea would have circumvented the obstacle.

We may never know. The sea now covers the evidence for coastal migrations, just as it does in so many different parts of the world where rising water levels have swamped and obliterated all traces of human habitation. An encampment with the discarded bones of deep-sea fish would prove an aquatic capability, even if the boats needed to catch them have long since perished. My admiration for the resourcefulness and resilience of our ancestors has grown to the extent that I find it hard to believe that their eventual discovery of the rich pastures of the United States south of the ice domes would have been thwarted by any of the temporary inconveniences that the ice ages might have put in their way. Where there is food, they will come.

Imagine the southerly progress of the first Americans, by land or by sea. Long before the rising temperature, the first thing they would have noticed was the light. The sun no longer sank to the horizon in the winter months, and it climbed higher into the sky in the middle of the day. Eventually they would have reached the southern limit of the ice and seen, stretching out before them, a land of wide tundra and rolling hills. Rivers swelled by ice melt gushed from under the glacial fringe, carrying frothing water gray with the suspended particles of tortured rock ground down to dust. Away to the south the rivers carried the pulverized rocks of Canada, until they settled out onto the flatlands of the Midwest, the waters flowing slowly now as the young Mississippi spewed out, exhausted, into the Gulf of Mexico.

The land was empty of humans but not of life. The animals were big, much bigger than they are today. The wild menagerie that greeted the pioneers who first encountered the rolling country beyond the ice contained a fair share of travelers who had, like themselves, arrived from distant lands. Three types of ancient elephant, all now extinct, had begun their own journey from Africa to the New World at least four million years before, following the same route as the humans who trekked through Central Asia and across to Alaska during one of the periodic openings of the causeway. The solitary forest-dwelling mastodon, the more sociable open-country mammoth, and its heavily insulated cousin, the woolly mammoth, were even larger than today's African elephant, and all of them reached America. Also much larger than its modern equivalent, the Ice Age *Bison antiquus* roamed over the entire continent. Musk oxen were the cattle of the frozen tundra, much as they are today, and like the bison, they had traveled across from Asia.

Some species that, you might have thought, would also have benefited from intercontinental travel stayed firmly put on the Asia side of Beringia. Both the yak and the woolly rhinoceros, though plentiful in Asia and certainly well adapted to the extreme cold, never made it to America. From what I have written so far you would be forgiven for thinking that the Siberian-Alaskan highway only took one-way traffic, but there were many species that originated in America and then emigrated to Asia and beyond. Both horses and camels first evolved in America from where they crossed Beringia eventually to become the two-humped Bactrian camels of Mongolia's Gobi Desert, and the one-humped dromedaries of the Arabian Desert and the Sahara. While camels survive in South America as llamas, alpacas, and the vicuñas of the Andes, wild horses became extinct in America and only re-appeared with the Spanish conquistadores in the sixteenth century.

Travel between North and South America only became possible between three and four million years ago, when the Isthmus of Panama rose above the waves thanks to the collision of two tectonic plates that thrust the seabed that connected an island chain upward and out into the open air. Until the two continents were linked, the evolution of South American mammals had taken an entirely different course from that of their cousins in most of the rest of the world. These were the marsupials, whose tiny embryos are forced early from the womb and suckled in a warm pouch. This is the fundamental difference between marsupial and placental mammals like ourselves, who emerge from the womb much more fully formed. Both independent evolutionary pathways produced very much the same type of mammal—herbivores, predatory carnivores, and so on—in either marsupial or placental form, but when the two were able to advance into the other's territory it was the placental versions that usually triumphed. The cat-like marsupial predators of South America were replaced by the placental ancestors of the jaguar. In Australasia, the only other place where isolation allowed marsupials to flourish, placental dingos replaced marsupial wolves (the last specimen nicknamed "Benjamin," passed into history in Hobart Zoo, in Tasmania, on September 7, 1936). Even so, some marsupials had crossed into North America and survived to greet the new human arrivals. None was more impressive than the giant ground sloth, which reached upward of eighteen feet at the shoulder.

All these animals were hunted by the first Americans. But there were very substantial dangers awaiting the new human arrivals: animals that would have come to know the taste of human flesh. Lions, larger and stockier than today's African and Asian survivors, were to be found all over America, both South and North, along with their fearsome cousin the *Smilodon*, known more descriptively, though inaccurately, as the saber-toothed tiger. But even these dangerous predators posed less of a threat to the human newcomers than did the quintessential distillation of fear and danger, *Arctos simus*, better known as the short-faced bear. These were not the ambling ground-hugging grizzlies, but long-legged bears built for speed. They were as tall as a moose and towered to fifteen feet when rearing on two legs. They could move as fast as the horses and bison that they hunted, and though a grizzly can put on a good spurt in pursuit of a wounded elk, the short-faced bear was a Bentley Continental to the grizzly's Nissan Micra.

How large a part the first Americans played in the extinction of the indigenous megafauna is open to question. Certainly, from the evidence of the fossil record, many species died out soon after they arrived, although a direct causal link is impossible to prove. What we can be sure of is that whether they came by land or sea, the ancestors of today's Native Americans overcame enormous challenges to settle the new continent. It is too easy to forget how resilient and courageous all our ancestors, wherever they lived, must have been to survive and to pass on their DNA to us, their descendants.

4

The Mystery of Cluster X

While the genetic evidence that we have covered so far does not agree with every aspect of traditional academic thinking about the origin of the first Americans, it does at least confirm that they arrived from Asia. Of the four predominant mitochondrial clusters found among Native Americans, all have their closest matches in Asia, even if three (A, C, and D) originated in Siberia and one (B) farther south in Taiwan and China. However, several years after the discovery of the four main clusters, signs of a fifth were found among some American Indian tribes. In all genetic surveys that I have been involved with, there are always one or two individuals who have mDNA sequences that appear to be completely out of place for their surroundings. A Polynesian sequence in Edinburgh or a Korean mDNA in Iceland are examples that I remember finding during the course of my own research. I call these "accidentals," borrowing a phrase from the world of ornithology, where the term describes the freak appearance of a bird in unfamiliar surroundings—an American robin, *T. migratorius*, blown off course to the west coast of Ireland, or a Wandering Albatross joining a colony of gannets on the Shetland Islands off the north coast of Scotland—that sort of thing. However exotic the explanation for the human "accidentals"—the elopement of a Tahitian princess or a pirate raid in the South China Sea—they are incidental to the overall genetic pattern of a region.

It was only in 1997, six years after the Nuu-Chah-Nulth work defined the mDNA sequences of the four main clusters, that a number of rare and apparently sporadic sequences that had been found in Native Americans were recognized as belonging to a separate cluster in their own right. By then the alphabetical notation of clusters had moved on a long way from A, B, C, and D, with global coverage having almost reached the end of the alphabet. With a fitting air of mystery, the new cluster was called X. The catalyst was a study of the Ojibwa.[1] The Ojibwa are the third-largest Indian nation in America, surpassed in numbers only by the Navajo and the Cherokee. They live on the borders of the United States and Canada in the general area of the Great Lakes. The mDNA of a quarter of Ojibwa volunteers belonged to the newly recognized cluster, far too high a proportion to be explained away as "accidental." Cluster X reaches its highest frequency among the Ojibwa, but elsewhere members of the cluster are found among the Sioux of the Great Plains, the Yakima of Washington State, and the Navajo of the Southwest, where the frequency of cluster X reaches 4 percent. The genetic connection between Ojibwa and Navajo also mirrors their related languages and may be an echo of a southward migration of the *Dine* (the name the Navajo give to themselves) mentioned in their mythology. Cluster X is found only in North America and did not find its way south of the Mexican border. But where had cluster X come from originally?

There is no trace of any cluster X sequences in Siberia or Alaska, and only a single example in China. But when the researchers looked the other way, toward Europe, they found plenty of genetic matches. I had often come across cluster X in my work on European mitochondrial DNA, and even recruited it to be among the *Seven Daughters of Eve*, as the descendants of Xenia, whose name begins with the cluster letter.

Could this be, as the authors of the first paper on cluster X in America suggested, genetic evidence of a European origin for at least some of the first Americans?[2] The radical suggestion was made in 1998, only two years after the discovery of the remains of a nine-thousand-year-old skeleton with, to some eyes, European features at Kennewick, Washington State. It was a discovery that, as we shall see, ignited an ugly and racially tinged argument about Native American origins.

I have reexamined the detail of the Ojibwa sequences and compared them with the now extensive European data to make a close comparison between the two. None of the Ojibwa sequences is exactly the same as their European cousins in the clan of Xenia, but they are not far off. The Ojibwa sequences do have their own mutations that are not seen in European Xenias, for instance the variant at 213, but the basic Xenia motif of 189 223 278 is there in both the Ojibwa and Europe. In fact one of the southeastern Ojibwa is a very close match indeed, with only a single change, at position 193, when compared with several Europeans.

Taken at face value, and remembering that unlike the other Native American mitochondrial clusters, X is not found in Siberia and eastern Asia, the evidence from the Ojibwa certainly suggests to me that they may well have a European origin. The difficulty is that this conclusion flies in the face of everything we think we know about Native American origins. Perhaps because of this, the enthusiasm for an early European arrival based on the evidence of cluster X waned among geneticists when the European and American sequences were found to be rather different in detail. But in my view that is only to be expected after a long period of isolation. The most recent genetic date estimate for cluster X in America is 15,800 years, so even with the inherent uncertainties of genetic dating, this is comparable with the age of other Native American clusters. Wherever it came from, from this evidence cluster X has been in America for a very long time. A European origin for cluster X raises all sorts of other questions.

The first of these is pretty obvious: If the ancestors of cluster X Ojibwa came from Europe rather than Asia, how did they get there? Second, as cluster X is only a minor cluster in Europe, averaging 5 percent, would it not be very strange that it was an ancestor from this clan who came to America rather than one of the others? But these paradoxes are no more difficult to explain than the alternative—that cluster X came from Asia. Here we would be asked to believe that members of cluster X joined with others in the journey from Siberia and Alaska, even though there are no signs of that cluster in northeast Asia today. While it is conceivable that cluster X became extinct in Asia at some point after the ancestors of Native Americans had left, it doesn't seem very likely to me. Either way there are big problems in coming up with an explanation.

Although we have a genetic date for Native American cluster X that is in the same range as the other four clusters, is there any direct evidence for its early presence in America, which would at least establish the cluster as one of the genuine founders? For this we need to look to the rather sparse evidence from DNA recovered from ancient human remains. As we already know, archaeological sites containing human remains are few and far between in America, and of these only a handful have yielded credibly ancient DNA sequences.

So far the most productive site to have surrendered the DNA of its ancient occupants is a burial mound at Norris Farms in central Illinois, overlooking the Illinois River valley. The name of the site comes from the corporation that owned the land on which the mound was discovered and immediately excavated in 1984. The grave goods identified the site as a burial ground for the Oneota, who had evidently used it for only a short time and comparatively recently. Carbon dating showed that the Norris Farms site is only about seven hundred years old, predating the arrival of the first French explorers to reach that part of America by only three hundred years. Although this was a dry site, and so not protected by anaerobic waterlogging, it was slightly alkaline, which helped to preserve the bones in good condition. And, as scientists from the University of Pennsylvania discovered, the DNA was in good shape, too. To the credit of the investigators, DNA was recovered from 108 of the 260 excavated skeletons, mainly from the rib bones.[3]

I have had a close look at the published Norris Farm sequences, and although they are from a comparatively recent site, it was reassuring to find that twenty-three of the twenty-five different mDNA sequences recovered from the bones were easily assigned to one of the four basic American clusters. That is not to say that all the sequences had been seen before in living Native Americans. About half had not, but they shared the core characteristics and could be placed in clusters A–D without any difficulty. That left two sequences that did not belong to the classic American clusters. One sequence had the motif 189 270, which is very common in Europe and lies at the heart of cluster U5 (the clan of Ursula, another of the *Seven Daughters of Eve*). The mitochondrial cluster that makes up this clan is more than forty thousand years old in Europe, and its arrival there coincides with the first appearance of anatomically modern humans in the European fossil record. There is no doubting the European credentials of this clan, or of the mDNA sequence recovered from the remains excavated at Norris Farms.

The other sequence that does not fit with the four American clusters was recovered from two other skeletons from the site, and has the core motif of cluster X plus two other mutations, at 227 and 357. While it is conceivable that the Ursula sequence is the result of contamination of the sample by DNA from a modern European, the same explanation is unlikely for the cluster X sequence from Norris Farms. First, it was found in two separate individuals, and second, it does not exactly match any modern Europeans. From this I think we can be confident that cluster X was in America at the time of Norris Farms, well before Europeans arrived in large numbers.

The ancient DNA evidence for an earlier arrival of cluster X rests on material recovered from a very unusual site at Windover, Florida. Windover lies near the Indian River lagoon in Brevard County, not far from Titusville and about twenty five miles from the NASA launch complex at Cape Canaveral. Windover came to light, as so often, just by chance when a digger operator, who had been hired to build a new road across the pond for a housing development, saw human skulls looking back at him from his bucket. He reported this, and after the sheriff had taken a look at them and declared that they were not recent burials, the developer, to his credit, halted construction and even helped to fund the archaeological excavation. One of the first tasks was to find a carbon date for the remains, which came back at eight thousand years.

Eight thousand years ago Windover was a woody marsh that was used as a regular burial site by early Indians. What makes this site exceptional is that while most bogs are acidic, the pH at Windover is comparatively alkaline, thanks to the buffering effect of high concentrations of calcium and magnesium carbonates that have leached out of a layer of snail shells lying beneath the bottom of the pond. Even weak acid destroys both DNA itself and the bone mineral that protects it, which is why bodies recovered from acid peat bogs resemble empty leather sacks. The skin is intact but the bones have dissolved, and so has the DNA. Windover was excavated in the mid 1980s, and a total of 177 bodies recovered along with carved bone and wood artifacts. About half of the bodies still had intact skulls, which, when opened, revealed a dark sticky mess. Examined under a microscope, the amorphous goo was unmistakably and very surprisingly identified as the remains of human brain. The preservation was so spectacular because the bodies, which had been laid to rest in a flexed position under about three feet of water, remained in a permanently waterlogged, anaerobic, and alkaline environment. Whether or not the Windover bodies were deliberately placed to maximize their preservation, the unusual conditions not only preserved the DNA but also stopped bacterial and fungal decay in its tracks.

I have also had a close look at the Windover data. Unlike the Norris Farms sequences, which are overwhelmingly drawn from the four known Native American clusters, the Windover sequences are highly variable and contain only one that is anywhere near a Native American sequence in cluster A.[4] The rest either have no matches that I can find or they are European, so I think considerable doubt surrounds the reliability of these data. Nonetheless one of them has the core cluster X sequence motif of 223 278, which if genuine, would put a lower limit of eight thousand years on the antiquity of the cluster in America. Even though the investigators were first-rate scientists, these sequences were recovered in the very early days of ancient DNA research, before anyone appreciated the magnitude of the contamination issues. Now that techniques have improved a great deal, it would, in my view, be well worth having another look at Windover. If cluster X sequences were found along with the other familiar Native American clusters, as at Norris Farms, this would be sure evidence that the clan was in America at least eight thousand years ago and, if the genetic date for the cluster of 15,800 years is to be taken seriously, considerably earlier.

At that time the Great Ice Age was near its maximum, so how could cluster X have reached America from Europe? If it were possible for cluster B to travel by boat around the margins of the Pacific about then, could the same be true of the Atlantic? Unlike the Pacific, there is no continuous, or almost continuous, land coastline to follow. However, in the depths of the last Ice Age and again during the thirteen-hundred-year-long Younger Dryas "cold snap" that ended just over eleven thousand years ago, much of the North Atlantic was frozen solid. A voyage around the virtual coastline of sea ice that coated the North Atlantic at the time does not seem completely out of the question. Then as now the ice was home to seals and other mammals, and fish were abundant. The leads in the pack ice could have given shelter from the worst of the storms and the northern summer would mean that much of the journey could be undertaken in daylight. In many ways a sea voyage around the ice coast of the North Atlantic would be not so very different from the conditions experienced by the ancestors of cluster B Americans on their voyages around the Pacific. As we always tend to underestimate the resilience and achievements of our ancestors, I think it just might have been possible.

Such a voyage was completed by the adventurer and novelist Tim Severin in 1976, when, with a crew of three, he left the west coast of Ireland in a hide-covered curragh and eventually arrived in Newfoundland. Earlier, in 1970, Thor Heyerdahl left the coast of Morocco, in North Africa, in the reed boat *Ra II* and reached Barbados. Both of these heroic voyages show that a transatlantic crossing is possible using only primitive craft, but like Heyerdahl's earlier expedition in *Kon-Tiki*, designed to demonstrate that Polynesia could have been settled from South America. But just because such a feat is possible doesn't prove that it actually happened. And for *Kon-Tiki* and the origin of the Polynesians, we know from the genetics that it did not.[5]

A sea voyage can also introduce a narrow genetic bottleneck. Only the people who get on the boat and survive the journey will leave descendants. You can see this effect very clearly in the Pacific, where the whole indigenous population of Polynesia appears to descend from just three women, who could very well have fitted onto a single craft at some point in the voyage. If something similar were to have happened in the early transatlantic ancestors in cluster X, this would help to explain why it is only this comparatively rare European cluster that reached America. A single boat with one woman in the clan of Xenia on board would have been sufficient. Any men who accompanied her would leave no mDNA trace.

The notion of a European origin for Native Americans is not new. During the seventeenth century it was a widely held view by many early colonists, including William Penn, the founder of Pennsylvania, that they were the descendants of the ten lost tribes of Israel. These were the remnants of ancient tribes that were dispersed following the destruction of the kingdom of Israel by the Assyrians about 720 BC. The Jewish origin of Native Americans is still a central belief of Mormonism. Other theories with popular support at various times in the past have seen Native Americans being descended from Phoenicians, Egyptians, Greeks, Romans, and even the Welsh, as well as, predictably, survivors of the lost world of Atlantis.

Shortly before the identification of cluster X among the Ojibwa, a chance discovery thrust the question of a European origin for Native Americans into the headlines once again. While human remains of early Americans are extremely rare, none was greeted with greater interest than the skeleton found in the Columbia River near the town of Kennewick in southern Washington State. On the afternoon of Sunday, July 28, 1996, two college students, Will Thomas and David Deacy, were sitting on the bank of the Columbia watching the thirty-first annual Tri-Cities hydroplane races. Hydroplanes are powerful and thrilling machines whose thrust lifts the hulls out of the water, and the "boats" almost fly across the surface. Even though there are still 250 miles to go before the Columbia empties into the Pacific at Portland, Oregon, the river is easily wide enough to accommodate such speeding craft.

Thomas and Deacy had positioned themselves on the south shore near the Columbia Park Golf Course. As Thomas waded into the shallows to get a better view of the race, his foot hit something hard and round. Reaching down, he pulled it out of the water and, in his own words, "saw teeth". It was a human skull. Not wanting to miss the afternoon's final race, the pair hid the skull in the bushes, returning later with a bucket in which they handed their grisly find over to the police. Their first thought was that the young men had stumbled on the victim of a fatal accident, or even a murder, and so they notified the local coroner. This is very reminiscent of the discovery of the "Iceman" in the Italian Alps in 1991, where the body was at first mistaken for that of a missing climber and its five-thousand-year-old antiquity was not recognized until much later.

Unsurprisingly the find was not covered in the U.S. papers, with the nation's attention on the pipe-bomb that exploded at the Summer Olympics in Atlanta the day before. Very little else of any lasting significance happened on July 28, 1996. The American papers reported a neighborhood shooting in Anaheim, California, a rousing "we must all pull together" speech by President Clinton, the conviction of a manufacturer for supplying substandard bolts for the space shuttle, and the publication of a new book by a *New York Observer* columnist. All but the last are long forgotten. The author was Candace Bushnell, and the book was *Sex and the City*, which has little, if any, genetic significance to the story unfolding on the banks of the Columbia River.

Sensing that this was no ordinary skull, the Kennewick coroner called in an anthropologist, James Chatters, who made the first excavation of the riverbank and recovered, over the course of the next few weeks, an almost complete skeleton. Its ancient pedigree was soon confirmed when Chatters found a stone spear point embedded in the hip. A more precise age for the remains of "Kennewick Man," as he was immediately dubbed, came from radiocarbon dating that returned a date of 9,300 years.[6] This was two thousand years younger than Clovis, but still a very old human by American standards. But what really catapulted Kennewick Man to a find of national and then international importance was a comment made by Chatters himself at a news conference. He claimed that the skull did not physically resemble any modern Native American. Instead Chatters said he thought that it looked much more like a European skull.

That comment, even though later denied by Chatters himself, immediately ignited a furor that threatened to undermine the widely accepted view that it was the ancestors of today's Native Americans who were the first people to settle the continent. The European connection was only enhanced by a facial reconstruction of Kennewick Man, which bore an uncanny resemblance to the English actor Patrick Stewart, best known internationally as *Star Trek*'s Jean-Luc Picard, commander of the starship *Enterprise* and successor to the legendary captain James T. Kirk.

To some the notion that America had already been reached from Europe thousands of years before Columbus was seductive. A few European Americans latched on to Kennewick Man as if he were one of their own ancestors who had discovered America. If this were true, then what did that say about Native American assumptions of priority that, after all, were the basis of land claims? The reasoning was perverted, of course, because even if some Europeans had arrived in America thousands of years ago, which I think they may well have done, they would have been rapidly subsumed, to be counted among the ancestors of today's indigenous people.

There ensued a vociferous custody battle for the remains, for surely whoever owned the body owned the myth. First in were the Asatru Folk Assembly, a quasi-religious group of Norwegian Americans based in Nevada City, California, and an offshoot of the Viking Brotherhood.

They petitioned the U.S. District Court in Portland, Oregon, to prevent the remains being handed to local Native Americans for burial. At about the same time eight anthropologists began a separate court action to prevent the remains being reburied, at least until a proper scientific investigation had been completed. The first Native Americans to stake a claim to the remains were the Umatilla, who wanted to rebury the remains according to tribal tradition.

The relevant statute here is the Native American Graves Protection and Repatriation Act (NAGPRA), passed in 1990. The basis for a claim is that a tribe may have custody if it can establish a kinship or cultural link to the remains. For the Umatilla that did not present a problem since, according to their own creation myths, their ancestors had been living in the region since the beginning of time. This claim had an additional facet of argument: Were the courts, and by implication the U.S. government, to reject their claim, it would be tantamount to a rejection of their religious beliefs, in contravention of the First Amendment. Three other tribes, the Colville, Yakima, and Nez Perce, followed the Umatilla with similar claims to what was left of Kennewick Man.

While the legal battles were being fought in the courtrooms, the remains themselves were given into the care of the Burke Museum in Seattle, from where scientists were allowed to conduct some limited investigations. The skeleton was that of a man around five feet eight inches tall who had been between forty and fifty years old when he died. The spear point buried in his hip was of the Cascade Point style, which corresponded nicely with the radiocarbon date of 9,600 years. Surprisingly, however, this could not have been what killed him as it was encased by new bone growth. But what everyone wanted to know was what his DNA was like, for that, it was thought, was the key to deciding if he really was a European.

Although the legal hurdles to the scientific study of Kennewick Man were eventually cleared in 2005, no positive genetic findings have yet been published. Looking in detail at the earlier attempts to recover DNA in 1996, I have my doubts that there is any left. Just ten days after James

Chatters assembled the almost complete skeleton and before the legal custody battle broke out, he contacted Frederika Kaestle, a graduate student
at the University of California Davis, and asked her to try to recover DNA
from the specimen. During early October of that year Kaestle took the
bone samples through the various steps needed to extract DNA, designing
her experiments around the mutations that defined the American mDNA
clusters. She found nothing conclusive, which was especially unfortunate
from a scientific point of view as, on October 19, UC Davis received a letter
from the U.S. Corps of Engineers, which owned the Columbia riverbank at
Kennewick, demanding that all DNA tests be discontinued. The university
decided, reluctantly, to comply, handing over the unused samples to the
Burke Museum to be curated along with the rest of the remains.

By spring 2000 there had been a change of heart at the Department of
the Interior, and the bone samples were retrieved and distributed to UC
Davis and two other centers, one of which was Dr. Kaestle's new laboratory at Yale. Reading the detailed description of their attempts to get out
some authentic DNA, I am reminded of the sheer frustration of trying to
do this with any but the best-preserved material. Following a very sensible
protocol, they duplicated DNA extractions as standard, but the duplicates
only rarely gave the same result. Positive controls were negative, and negative controls positive. Students in my lab doing this kind of ancient DNA
work with poorly preserved remains were sometimes reduced to tears.
Catching the same sense of frustration in the reports on Kennewick Man,
I was very sympathetic when I discovered that the only DNA that any of
the investigators managed to extract reproducibly was their own.[7]

Kennewick Man remains an enigma and, like the Windover brains,
would certainly benefit from a reanalysis using the improved DNA techniques available today. However, the circumstances have become more
unfavorable for an unbiased scientific investigation because of the misguided claims that have already surfaced. I have no sympathy for the
claims of the Asatru Folk Assembly and the implication that finding a
typically European DNA in Kennewick Man would invalidate Native

American claims to be descended from the first Americans. For one thing, no DNA result would lend scientific support to the Asatru claim since I do think it quite likely, for reasons I have explained, that there is some ancient European ancestry among Native Americans. But the legal tussles surrounding Kennewick Man also highlighted the beliefs that Native Americans have of their own ancestral origins. This, and the Native American view of genetic ancestry research, is something we will consider in depth in a later chapter, but to give you a taste of the problem—and it is a problem—let me recount the case of the Havasupai.

I was aware that the results of DNA testing had the potential to conflict with tribal oral histories, but I have never encountered this as a particular difficulty in my own research in Polynesia or in Britain. Polynesians were generally curious to discover the location of their mythical homeland, Havai'iki, and eager to help with my genetic research, which might throw some light on it. Equally, British men and women needed little persuasion to volunteer their DNA to help discover whether the ancestors of today's Britons were Saxons, Vikings, or Celts. I always thought that I would soon become aware of any potential conflict when I visited an area and talked to the volunteers. If they didn't want me to look into their genetic ancestry, then neither did I. However, it is pretty clear that not all genetic research has been done in this way in the past—especially when there was a rush to make new discoveries before anyone else could publish, as in the case of Native Americans. Tellingly, the influential paper on the Nuu-Chah-Nulth that we covered in the previous chapter was carried out without consent on samples collected earlier for a different purpose. There was a temptation to sidestep the collection phase, which would soon have uncovered any sensitivities, and use DNA from previous collections. I don't think this was done in deliberate contravention of the wishes of the donors, but more because the investigators were mainly European, among whom genetic ancestry testing is largely uncontroversial. The principal author on the Nuu-Chah-Nulth paper later admitted, "The way people operated at the time, it didn't cross anyone's mind—we didn't mean to be evil and we are more careful now."[8] This was a bad misjudgment in the case of Native

Americans, as the protracted legal dispute between the Havasupai Indians and the University of Arizona showed very clearly. Matters came to a head in April 2010 with the *New York Times* putting the out-of-court settlement on the front page under the headline "Tribe Wins Fight to Limit Research of Its DNA." The accompanying photograph of a waterfall cascading from red cliffs into a turquoise pool immediately placed the tribal home of the Havasupai in the region of the Grand Canyon.

To explain the background to the case, American Indians have among the highest rates of type 2 diabetes in the world. The best known of these are the *Akimel O'odham*, or Pima Indians, who live alongside the Gila River in southern Arizona and have been studied intensively since 1965 by an offshoot of the National Institutes of Health in nearby Phoenix. The prevalence is frightening, with half of those over thirty-five years having the condition.

I don't view this form of diabetes as a disease in the usual way because I subscribe to the "thrifty gene" explanation first put forward by the geneticist Jim Neel in 1962.[9] This explains diabetes as a genetic condition that has advantages when food is short and only becomes a problem when high-carbohydrate foods become abundant in the diet. When times are hard, as they have been for the ancestors of all Native Americans, as we have seen, the individuals carrying the thrifty genes are favored by natural selection because, with their more efficient metabolism, they can survive on less food than their contemporaries. But when food, especially carbohydrates, which were virtually unknown to our hunter-gatherer ancestors, becomes more plentiful, the thrifty genes are a distinct disadvantage. Unaware that starvation is now no longer a threat, these genes continue to direct the body to store fat in preparation for the winter shortages that never come. I have seen the same pattern in Polynesia, where thrifty genes helped the original settlers survive the long voyages across the vast Pacific but now make the modern-day descendants of these intrepid navigators prone to obesity and diabetes when consuming a high-carbohydrate diet. The same is true, I am convinced, in Europeans, whose hunter-gatherer ancestors also faced the

threat of starvation every winter. As we are all aware, the incidence of
type 2 diabetes is on the increase in the well-fed West, and there is every
reason to find out the genetic cause. I would not go so far as to say that
it is for this reason alone that diabetes among the Pima has received so
much research attention: It is a very real problem for them, not with-
standing the universal interest in the outcome.

Although the Pima are by now the textbook example of diabetes
among indigenous peoples, and I have included them in my genetics lec-
tures to medical students, it was only natural to be curious whether the
same thrifty genes, whatever they were, were to be found in other Indian
tribes where there was also a high incidence of diabetes. That is where the
Havasupai first became involved. According to the *New York Times* article, it was the Havasupai themselves who asked an Arizona State University anthropologist who had been working with the tribe for several years
whether he knew of anyone who could help discover the cause of the condition that was forcing tribal members to have limb amputations and to
leave the canyon for kidney dialysis. As a direct result of this request, Dr.
Therese Markow began working with the Havasupai shortly afterward,
aware that finding a common genetic risk shared by the Pima might help
to pinpoint the diabetes genes. She began to collect samples in 1990, bas-
ing herself at the tribal village of Supai, deep in the Grand Canyon.

Over the next four years Dr. Markow collected blood samples from
about a hundred tribal members, who, having had the aims of the project
explained, signed a broad consent to their samples being used "to study
the causes of behavioral/medical disorders." As she was also interested
in the genetics of schizophrenia, the consent was broader than diabetes alone. Back at Arizona State University, DNA was prepared from the
blood, and Dr. Markow's team started the long process of looking for
genes among the Havasupai that might be implicated not only in diabetes but also in schizophrenia and alcoholism. Not unusually for the
1990s she did not discover any strongly associated genes for any of these
conditions, and the unused DNA remained in the lab freezer.

Several years later, by which time the techniques for locating susceptibility genes had improved, Dr. Markow assigned a student to have another go at the Havasupai samples. It was when the student had finished the work and was about to present it for public examination in 2003 that the trouble erupted. Carletta Tilousi, one of the Havasupai who had volunteered her DNA back in the 1990s, was invited to attend the public examination. According to her account, what she heard bore no relation to the diabetes project she thought she had been invited to join. She asked bluntly whether the student had permission to use Havasupai blood for her research. At this point the presentation was halted, and subsequently the offending chapter was removed from the dissertation and the university launched an investigation. This revealed that the Havasupai DNA samples had been used for several projects unconnected with the original diabetes investigation.

Two of these caused particular offense to the Havasupai. One was that the genetic data were used to calculate the degree of inbreeding within the tribe. The subject of inbreeding among the Havasupai and other Indian tribes is extremely sensitive, as indicated by their belief that if it is discovered, a relative will die. Second, and the reason I raise it here, is that the Havasupai samples were also included in research that explored the ancestral origin of Native Americans. As was already evident in the claim by the Umatilla for custody of the Kennewick remains, they, like the Havasupai, do not believe that their ancestors came from Siberia or Europe or anywhere else. In their tradition, which is shared by many if not most Native American nations and tribes, their ancestors have been living where their descendants are now living since the beginning of time. As we shall see later, the Havasupai case illustrates the conflict between traditional beliefs and modern science, which has serious consequences for Native Americans and geneticists alike.

In his sobering 2010 presidential address to the American Society of Human Genetics, Dr. Roderick R. McInnes condemned the carelessness and cultural ignorance of geneticists that led to the mistakes of past decades of research on indigenous people.[10] There can be no more excuses after such a high-level intervention.

5

The Europeans

n complete contrast to the suspicion and hostility with which, for understandable reasons, Native Americans now regard genetic research, European Americans have welcomed such investigations with open arms.
But before I explore the results of this enthusiasm, let me just briefly go
over the events that brought so many Europeans to America in the first
place. Though these will be more than familiar to most readers, I do not
want to assume that everyone has them at their fingertips.

In the modern era, by which I mean since the last ice age, the first
European known to have visited America was the Norseman Leif Erikson in AD 1002 or thereabouts. Born in Iceland, Erikson was the son
of the Norwegian explorer and outlaw Erik Thorvaldsson, who founded
the first settlements in Greenland. Erikson sold plots of land to gullible
Icelanders who fell for his description of a green and fertile land, which
though wildly misleading, has stuck in the name. He was also a born
explorer and set off north and west from Greenland in an oceangoing
longship to look for new land that a friend of his had reported seeing.
He soon came across Baffin Island, then headed south, sighting the coast

The *Mayflower* in heavy seas.

of Labrador, and finally landing on the tip of Newfoundland. Here he established a temporary settlement, called Vinland, after the wild vines he found growing there. Despite the hospitable surroundings, and rivers full of salmon, Erikson sailed back to Iceland, never to return. Being a great admirer of the Viking spirit of adventure, I find it very hard to believe that news of this voyage did not spread and was not followed up, if not by Erikson then by someone else. But apparently not, and it was another five hundred years before another European was to step ashore. Nevertheless Leif Erikson's exploits are celebrated in the United States every October 9 with a special commemorative day, although the great majority of Americans today are not aware of it.

By some accounts the story of Erikson's discovery was still well-enough known when Christopher Columbus visited Iceland in 1477, fifteen years before his first transatlantic voyage in 1492, hoping to reach India and encouraged to do so by a severe underestimate of the world's circumference. Instead he landed first in the Bahamas and then in Cuba, before finally running aground on the island of Hispaniola, now shared between Haiti and the Dominican Republic. In all, Columbus made four transatlantic voyages between 1492 and 1504, all of them sponsored by the Spanish crown, exploring more Caribbean islands and the coast of Central America but never sighting the North American mainland. For the next hundred years it was Spain and Portugal that were behind most new European colonies, though they concentrated more on South and Central America than the North. They soon eliminated opposition in Mexico, when they defeated the Aztec, and in Peru, where they destroyed the empire of the Inca. In North America the Spaniards built outposts in Florida in 1565 and then in Arizona and New Mexico, establishing Santa Fe in 1610. By the middle of the eighteenth century they had founded colonies along the coast of California, centered on Monterey.

Although the Spaniards were the first Europeans to settle in America, other European powers were not far behind. France, Holland, and even Sweden founded colonies on American soil, but it was the British who, after a protracted start, eventually became the predominant European influence,

particularly along the eastern seaboard. The first English colony was established in August 1585 on Roanoke Island, among the shoals that guard the
entrance to Albemarle Bay, North Carolina. Five years later, in 1590, a relief
ship found the colony deserted with no sign of the 120 inhabitants. There
was no evidence of a struggle, and the buildings had been deliberately dis-
mantled, as if the colonists had decided to relocate. They were never seen
again, and their fate remains a mystery to this day.

The first permanent English colony was begun at Jamestown, Virginia,
in May 1607 on the James River, forty miles upstream from Chesapeake
Bay. The site was swampy, regularly inundated by brackish sea water, and
infested with mosquitoes. Eighty percent of the colonists died within two
years. Yet Jamestown survived—barely—and became the center of the
Virginia Colony, with the population growing to forty thousand by 1670.
Its survival was only marginal in the early days (and Bostonians claim
Jamestown was formally abandoned for a short while, thereby ceding the
title of first *permanent* settlement to their own city). All the expected
crops of sugarcane, oranges, and lemons failed, and it was only the introduction, in 1612, of tobacco, which thrived in the damp heat, that saved
the colony. By the 1650s the Virginia Colony and the neighboring colony
of Maryland at the north end of Chesapeake Bay were exporting annu-
ally five million pounds of tobacco back to Europe.

Farther north the English Puritan colonies of New England also got
off to a shaky start following the arrival of the *Mayflower* and its complement of a hundred Pilgrims at Plymouth in Massachusetts Bay in
November 1620. Famously they survived the winter thanks only to the
help they received from the local Wampanoag Indians. Unlike the commercial motivations of the Virginia Colony, the Plymouth Colony was
started to escape religious intolerance in England. More colonies followed. In 1630 John Winthrop arrived on the *Arabella* with a royal char-
ter to establish the Massachusetts Bay Colony, centered on Boston, with
himself as governor. Others, like Rhode Island, were started as result
of fierce internal theological disagreements. By 1640 there were about
twenty thousand colonists living throughout New England.

The population grew steadily but not spectacularly, so that by the time of the Declaration of Independence in 1776, somewhere between five hundred thousand and a million people, mainly from Britain and northwest Europe, had crossed the Atlantic to begin a new life in America. Many arrived as indentured servants, working for several years to repay their employers the cost of their passage. When they had worked off this debt they were, in theory, granted land of their own, although this often did not materialize. The beginning of the nineteenth century saw a very large expansion in the number of immigrants from northern Europe following the Louisiana Purchase of 1803, when Thomas Jefferson engineered the acquisition of territories originally claimed by France west of the Mississippi for $15 million ($220 million today). The transcontinental expedition of 1804–6, undertaken at Jefferson's request by Meriwether Lewis and William Clark, set off from St. Louis to explore the new territories along the Missouri and farther west in what were to become Washington and Oregon. As well as its scientific and geographical mandate, one objective of the expedition was to establish a claim over these northwest territories before the British, who were operating out of Canada.

Through statutes like the Homestead Act of 1862, successive US governments encouraged large-scale immigration and settlement of the West, partly to discourage the British. Coupled with industrial growth in the Northeast, this territorial expansion attracted more than twenty-five million Europeans, mainly from northern Europe, to emigrate to the United States by the end of the nineteenth century. At the turn of the twentieth century, Europeans were still arriving in large numbers, but mainly from southern and eastern Europe, and this trend continued until the Immigration Act of 1924, which attempted to limit immigration of Jews, Italians, and Slavs by imposing national quotas. European immigration slumped during the Great Depression, but the quotas were retained, eventually becoming one factor that prevented many Jews from escaping Nazi persecution by emigrating to the United States. National quotas were abolished in 1965, since when the great majority of immigrants have originated from outside Europe. Even so, a glance at the

population ancestry map of the United States following the 2000 census, which plots the population origin of the majority residents in each
county in each state, shows that Europe is still the ancestral origin for
most Americans. (See map in color insert section.)

The effects of large-scale immigration during the nineteenth century
are very clearly reflected on the map, with twenty of the forty-eight
states in the contiguous United States (that is, omitting Alaska and
Hawaii) drawing their largest population from Germany. Italian Americans predominate in New York State, Irish Americans are the largest in
Massachusetts and New Hampshire, while Britons are the majority origin only in Vermont, Connecticut, and, through their Mormon origins,
Utah. However, in Arkansas, Tennessee, Kentucky, and West Virginia
the largest number self-declared as "American," a category that no doubt
encompasses many with ultimately British roots. In four southwestern
states with historical ties to the early Spanish territories—California,
Arizona, New Mexico, and Texas—the greatest number declared Mexican origins. In seven southern states—Louisiana, Missouri, Alabama,
Georgia, North and South Carolina, and Virginia—the largest proportion declared themselves as African Americans.

   In no states were American Indians in the majority, but when the states
were subdivided into counties, they did make up the largest proportion of
the population in parts of Arizona and New Mexico, Oklahoma, South
Dakota, and Montana. These higher-resolution plots also showed that the
overall German dominance in the northern half of the United States was
punctuated by counties with their greatest proportions drawn from other
European countries. Norwegians dominate in parts of North Dakota and
Minnesota, while Finns are in the majority in some regions of Michigan
bordering Lake Superior. In a few counties within southern Michigan and
Iowa, Dutch Americans are in the majority, while Americans of French
origin predominate around their former colony in New Orleans. Perhaps unexpectedly, there is an Irish majority in two counties in southern
Washington State and around Butte, Montana.

Apart from the numerical dominance of European Americans in large swaths of the United States, the other, subtler feature of the map is that so many respondents in the 2000 census knew what their ancestral origin was, at least sufficiently to declare it as such on the census form. This reflects a key feature of European American ancestry, which is that almost everyone either knows what their European origins actually were or has a strong belief and attachment to them. In complete contrast to Native Americans, there is no general uncertainty surrounding the origins of European Americans and therefore no pressing need to recruit genetics to help sort these out.

Among the potential exceptions are the questions surrounding the fate of the vanished colonists from Roanoke Island; whether or not the original Viking explorers who landed in Newfoundland left any descendants; and last, whether other European expeditions to the New World predated Columbus and left genetic evidence behind. I have been asked to look into all three of these puzzles in the past, but I have not undertaken any of them because I think there is very little chance of a clear result. It is not that there is any difficulty in distinguishing Native American from European DNA. That is easy. The difficulty lies in the sheer numbers of European immigrants who followed on the heels of these early arrivals. This makes it, in my view, quite impossible to be sure that, for example, any Norse origin DNA found among the modern-day inhabitants of Newfoundland could be confidently attributed to Leif Erikson's men rather than to more recent Scandinavian immigration. The same applies to English DNA around Roanoke in North Carolina or to Portuguese genes in the Caribbean. That isn't to say it can never be done—the best route would be the recovery of DNA from well-preserved and well-dated human remains—but not at present by comparisons with the modern populations.

Nevertheless, while very few general mysteries surround the ancestral origins of European Americans, there is no limit to individual curiosity that, amplified by the importance attached to these origins reflected by the census returns, has given genetics plenty to do over the course of the last decade. When it comes to European Americans it is genealogy rather than anthropology that has gained by involving genetics. And while the research on the origins of Native Americans largely relied on mitochondrial DNA, as we have seen, genealogical research has involved the other important witness to the past, namely the Y chromosome. While mDNA tells the story of women, the Y chromosome is a chronicler of the behavior of men.

This intriguing piece of DNA resides in the cell nucleus rather than the cytoplasm and is one of the twenty-four different chromosomes that together make up the human genome. But, while it has some of the properties of its other chromosomal companions in the nucleus, it is, by its very nature, an outsider. It travels alone through the generations without exchanging DNA with any other chromosome. Only men have Y chromosomes, and the reason for this is both straightforward and utterly fascinating. After fertilization, all human embryos start off as female and, if nothing happens to change their development, are born as baby girls. However, the Y chromosome contains a single gene that, when activated at about six weeks after conception, diverts the embryo from a female to a male trajectory. How it does this is complex and not completely understood, but the end result is straightforward: An embryo with a Y chromosome turns into a boy, while an embryo without a Y chromosome continues developing into a girl.

When they set out to fertilize an egg, half of the sperm contain a Y chromosome and the other half do not. If the fertilizing sperm is one of the 50 percent with a Y chromosome, then the child is a boy and, conversely, if the fertilizing sperm does not have a Y chromosome, the baby will be a girl. This is why approximately half of babies are boys and half are girls. Thus, Y chromosomes are passed on exclusively from fathers to their sons.

While it is the Y chromosome that has caught the imagination of genealogists, many European Americans also have an interest in their maternal ancestry, which is tracked very well by mitochondrial DNA. As we have seen, there are five clans among Native Americans, four of them, Aiyana, Ina, Chochmingwu, and Djigonase, from Asia and one, Xenia, from Europe. Xenia is one of the seven predominant maternal clans my research team and I first uncovered in the late 1990s and which formed the basis for *The Seven Daughters of Eve*. The cluster date for Xenia in Europe we estimated to be twenty-five thousand years, but that was an uncalibrated date not taking account of the founding lineages. My colleagues Martin Richards and Vincent Macaulay then undertook the Herculean task of combing European and Middle East data for mDNA matches in all the clans that would identify likely founder sequences, after which they recalculated the cluster dates. We published the results in 2000, and it was a relief to find that they were largely in line with our earlier uncalibrated estimates and reinforced our conclusion that the majority of the clan mothers were in Europe, as Paleolithic hunter-gatherers, before the arrival of Neolithic farmers from the Middle East about ten thousand years ago.[1] You can find these calibrated cluster dates in the appendix. As usual with adjusted dates, they are younger than their uncalibrated equivalents because eliminating founder sequences reduces the average number of mutations in the remainder.

One difference is the replacement of one clan (Velda) by another (Ulrike) in the ranking of the seven most prolific European clans. This is due to the extended geographical range of the analysis that took in new results from eastern Europe that had not been available in our first analysis. Velda is very much concentrated in western Europe while Ulrike is more common farther east. Another difference is that we recognized a major branch of the Tara clan (T2) that had a considerably younger date than the age estimates of the clan as a whole. This branch is probably of Neolithic rather than Paleolithic origin and joins the clan of Jasmine in entering Europe with the introduction of agriculture.

All these clans are well represented among European Americans, exactly as expected. Moreover, clan frequencies in the United States are much the same as they are in the regional source population in Europe. Since the ancestors of all European Americans arrived within the last five hundred years, and many far more recently than that, there has been very little time for new mutations to accumulate among their descendants living today. Even if the ancestors of all European Americans had arrived with Columbus five hundred years ago, the average mutation rate of one change every twenty thousand years means that that only one American in forty (20,000/500) would have a sequence that differed from his or her European ancestors. For Americans in search of their maternal relatives in Europe this has the advantage of limiting their search to only those individuals with exactly matching mDNA sequences.

Like mDNA, the Y chromosome groups individuals into clusters with a common ancestor, the only difference being that these clusters are patrilineal rather than matrilineal. Although Y chromosome DNA (or yDNA from now on) is much longer than mDNA, one type of variation is the same for both: That is the differences in sequence introduced by faulty copying, whereby one base is substituted for another. However, the rate of mutation is about twenty times slower in yDNA because, as with the other nuclear chromosomes, there is a very efficient quality-control mechanism that identifies copying errors and eliminates most of them. Mitochondria don't have this error-checking mechanism, and so mutations have a better chance of getting through to the next generation. It was a real struggle to identify the yDNA sequence changes, but the hard work has been done, principally by Peter Underwood and his team at Stanford University, and we now have a good range of markers that are capable of distinguishing thousands of different Y chromosomes from one another. Because of their slow mutation rate these markers, which go by the acronym of SNPs (for "single nucleotide polymorphisms" pronounced "snips") have each probably changed only once during the course of human evolution and so are very useful for plotting out the overall Y-chromosome family tree. However, they are not very much use to genealogists interested in much more recent time frames.

Fortunately help was at hand because a completely different type of yDNA marker was discovered in the mid-1990s and looked as if it was tailor-made for genealogists. The relevant acronym here is VNTRs, standing for "variable numbers of tandem repeats," which sound more complicated than they really are. VNTRs are segments of DNA composed of repeating blocks of DNA sequence usually three to five bases long. They are very boring to read through, for example, AGTAGTAG-TAGTAGTAGTAGTAGTAGTAGT where the triplet AGT is repeated ten times. What makes them interesting is that the number of times the triplet is repeated varies between different Y chromosomes. It is relatively easy to measure the length of these variable segments and so work out how many repeats any Y chromosome contains. In this example, instead of ten repeats, there might be anywhere between eight and twelve on different Y chromosomes. This is an excellent way to distinguish chromosomes that would otherwise be impossible to tell apart using the SNP system. And because Y chromosomes are the outsiders of the cell nucleus and don't talk to or exchange DNA with their companions, they remain completely intact. This unsociable behavior has a tremendous advantage because it means that all the markers along the whole chromosome (well, barring the very ends, which we can ignore) are inherited together from one generation to the next. This in turn means that results from one marker can be combined with another and the effect is multiplied. So if there are five different repeats at one marker, like our example, and five at another, the number of possible and distinguishable combinations is not $5 + 5 = 10$ but $5 \times 5 = 25$. You can imagine how the number of different Y chromosomes that can be recognized by the VNTR system increases very quickly, so that, with ten markers each with five different lengths, the number of combinations is $5^{10} = 9.7$ million, and with twenty it reaches a staggering $5^{20} = 94$ trillion (where a trillion is a million million). As this figure greatly exceeds the world's entire male population of roughly 3.5 billion, it is plain to see that not all possible Y chromosome combinations actually exist. With this amount of variation available the scene was set for the revolution in genetic genealogy of the last decade.

6

The Genetic Genealogy Revolution

The speed and enthusiasm with which the American genealogy community has embraced genetics has been truly astounding. I vividly remember addressing a meeting of the New England Historical Genealogy Society, America's oldest, in 2001 when only a tiny minority in the audience knew much about DNA and hardly anyone had heard of mitochondria or Y chromosomes. Now knowledge is detailed and extensive, and genuine advances are being made, fueled by the curiosity of members of the public about their roots rather than by academics.

The very first scientific paper to feature Y chromosomes in any sort of genealogical connection concerned a man who has already made an appearance in this book, and will do so again. He is Thomas Jefferson, third president of the United States and principal author of the Declaration of Independence. Jefferson's wife, Martha, died in 1782 following the birth of her sixth child, Lucy Elizabeth. Some years later Jefferson and one of his slaves, Sally Hemings, who was the half-sister of his late wife, became lovers and he may have fathered a further six children by

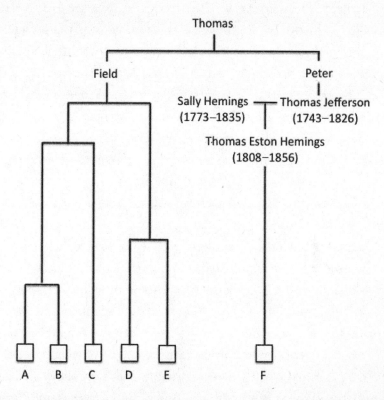

Figure 1. Patrilinear relatives and descendants of President Thomas Jefferson with matching Y-chromosome signatures. Lengths of vertical links are approximately proportional to the number of generations.

her including her last son, Thomas Eston, born in 1808. The story is a fascinating one for many reasons, not least because, given his political prominence and the unforgiving nature of the times, Jefferson denied it. The scandal, for that was what it was, rumbled on for the next two hundred years until 1998, when genetic evidence proved, beyond reasonable doubt, that it was true. The study that clinched the verdict compared the combinations of markers in the Y-chromosome signature from a direct patrilineal descendant of Thomas Eston Hemings with equivalent relatives of the president (Figure 1).[1] Thomas Jefferson did not have any surviving legitimate sons, so his Y chromosome had to be identified through a patrilineal relative, who in this case was the president's paternal uncle Field Jefferson. Both Field and Peter Jefferson, the president's father, had inherited the same Y chromosome from their father, Thomas. Therefore any patrilineal descendant of Field Jefferson would carry the same Y chromosome as the president. When the Y chromosomes of five such descendants of Field Jefferson (A–E) were compared with a direct descendant of Thomas Eston Hemings (F), they matched exactly. Moreover, the precise Y-chromosome signature was rare in the general population, making the match extremely significant. This conclusion has not gone down well with some Jefferson descendants, but it was the result of a decisive piece of work.

Although the Jefferson/Hemings case demonstrated the power of genetics to prove a genuine patrilineal connection, it did not make a general case for the use of the Y chromosome to follow surnames. In fact, quite the opposite. Because Thomas Eston Hemings was illegitimate, the son of a slave and her master, and carried his mother's name rather than the president's, it was a prime example of what geneticists call a nonpaternity event, where a surname does not follow the same line of descent as a Y chromosome. This could be because of illegitimacy, as here; adoption; a deliberate name change; or infidelity by the mother. Whenever this happens, the link between a surname and its Y chromosome is broken forever. It was the disruptive effect of nonpaternity on surname/Y-chromosome associations that persuaded those few geneticists who thought about such things at the time that the Y chromosome was unlikely to prove useful on a larger scale. I think we were swayed by the generally high rate of nonpaternity, sometimes as high as 10 percent, uncovered by conventional genetic fingerprinting among the modern population, whereas it was the historical rate that was more relevant. Only when I was curious to test another man with my surname, Sir Richard Sykes (at the time the chairman of the pharmaceutical giant Glaxo), to see if we were related, did the surprisingly high general correlation between surnames and Y chromosomes in men start to come to light.

After I found that our Y chromosomes matched and so we were likely descended from a common ancestor, I randomly recruited around one hundred other male Sykeses from the phone books of West Yorkshire, where the name is concentrated, and found that 60 percent of us also shared this same Y chromosome. I have written about this extensively in *Adam's Curse*, so I won't repeat myself here. Suffice it to say that the association must have lasted for eight hundred years, since, like many other English names, Sykes became hereditary during the thirteenth century. For that to be so, the nonpaternity rate must have been much lower than I or my colleagues had, rather lazily and certainly mistakenly, assumed. For the Sykes name this rate worked out at just over 1 percent per generation. For other surnames, like Dyson from the same Yorkshire valley, it was even less, with 95 percent of Dysons sharing the same Y-chromosome fingerprint. I was fortunate in that the shared Sykes Y chromosome was rare, rather like Jefferson's in that respect, as this made the matches very significant from a statistical point of view because they were very unlikely to have happened by chance. I think the effect was so striking, though, that even if the Sykes Y chromosome had been a common type, the fact that 60 percent of us shared it would have shown up as unusual.

Not being a genealogist, I was not really aware of the practical importance of this discovery, but that was to change very fast. Within days of the publication of the paper announcing the Sykes results, in April 2000, the lab was swamped by requests from members of the public asking us to help with their own genealogical research.[2] The scientific paper had a fair amount of press coverage in Britain, including an appearance on breakfast television where I remember sharing the sofa with the world chess champion, Garry Kasparov (and in complete contrast and basically setting the tone of the show, a man who could balance on a coconut using only his thumbs). In the United States there was an article in the *New York Times*. The following week the London *Times* carried a feature on the seven clan mothers I had identified and named, following the mitochondrial DNA research in Europe. That also triggered a huge response from the public, with requests for genetic tests. So we were very soon faced with a choice in the lab, and I called a meeting in the conference room to come to a decision.

We agreed that we could not, and should not, use our research time and money to test individual families purely for their own interest. Once the general principle of the surname/Y-chromosome link had been established with the twenty or so names we were then working on, we would not be able to justify much more work, except in special cases with their own intrinsic academic interest. Our choice was, then, whether to decline these requests from the public or to set up a proper mechanism for dealing with them. We debated this long and hard. One or two members of the lab were apprehensive and against any "commercialization," as they saw it. But in the end the argument that won the day was that we had by chance stumbled onto something that evidently appealed to a large number of people. To do nothing about it seemed to me and most of my colleagues to be tantamount to saying that we were only doing this research for the small audience of our scientific peers, not for the public. We agreed that we should respond, and that it was right to do so. We set up Oxford Ancestors that same week, hired a technician, and began taking orders for both Y-chromosome and mitochondrial DNA tests, the first company in the world to do so.

It took another year to set up an operationally independent commercial arrangement with separate staff and premises in a local business park. The company is still going strong and has recently celebrated its tenth anniversary. I am sure I was mistaken in thinking that if we didn't do something about the public demand, it would go unsatisfied—there are now plenty of other companies doing this sort of thing—but I don't regret doing so for a minute. Oxford Ancestors has helped tens of thousands of people "explore their genetic roots," as it says in the company logo. It has been a great experience seeing so many aspects of what people can discover from their DNA, and along the way I have encountered some absolutely fascinating reactions. As I write, a decade into the era of public accessibility, I estimate that almost a million people have had their DNA tested for either mitochondrial or Y-chromosome DNA, or both. A lot of this has been personally financed, as it should be, through private companies like Oxford Ancestors in the UK, Family Tree DNA and Relative Genetics in the United States, as well as through American companies specializing in African ancestry.

Though Oxford Ancestors began in my university genetics lab in response to an unanticipated demand, other companies had different primary motivations, even if all of them ended up doing much the same thing. In the case of Houston-based Family Tree DNA, it was the curiosity of its genealogist founder that provided the vital spark, as the ebullient Bennett Greenspan explained when I visited him at the company premises on the top floor of a commercial building on the outskirts of this thinly spread city. Greenspan had taken an interest in the work on the Y chromosomes of the hereditary Jewish priesthood, the Cohanim. Briefly, the male descendants of Aaron, the brother of Moses, were selected to serve as priests. Genetic corroboration of the Old Testament tradition came when a particular Y chromosome was found at much higher frequency among Ashkenazic Cohanim than in comparable non-Cohanim. Greenspan became aware of this when he read the *Wall Street Journal* report of the earlier *Nature* publication in July 1998.[3] He knew nothing about DNA, as he readily admits, but was already an experienced genealogist and, as a Jew, the Cohanim story caught his eye. The news also came at a time when he had just sold his photographic-products business, having correctly anticipated the effect of the digital camera revolution on the demand for film. So he was on the lookout for something new into which to channel his energies. His first visit was to the University of Arizona and the laboratory of Dr. Michael Hammer, one of the early pioneers of uncovering the genetic variation in the Y chromosome on which all subsequent studies depended. Although Hammer was not one of the members of the Cohanim project, which was carried out in England and Israel, he was the obvious man to speak to. And, in American terms, it wasn't that far from Houston to Tucson.

Bennett Greenspan had a particular project in mind when he called on Hammer. He was trying to see whether another man named Greenspan, from Argentina, was related to himself, and having been unable to find a paper trail connecting the two of them, he realized the potential of the Y chromosome to solve the issue. It was as he was leaving that Hammer said to him, "Someone really should turn this into a business"— which is exactly what Bennett Greenspan did. Thus, Family Tree DNA was born, opening for business within weeks of Oxford Ancestors' debut in 2000. As it turned out, he and the Argentinean Greenspan were not genetically related, and before you ask, no, Bennett still doesn't know if he is related to his namesake Alan, the former chairman of the Federal Reserve.

The third early genetic genealogy company also sprang from a university, but in a rather different way from Oxford Ancestors or Family Tree DNA. For many years Salt Lake City had been at the center of global efforts to map and sequence the human genome. The city's location and its connection to the Mormon Church with its deep interest in tracing family connections and unequaled historical records made it a natural place to explore the potential for genealogy of the new genetic knowledge. I had gotten to know Dr. Scott Woodward from Brigham Young University when we had both been working in the field of ancient DNA, so it was no surprise to find him leading the initiative as head of the Sorenson Genome Institute. I had planned to visit him, but a bush fire in Yellowstone National Park intervened, and I had to ask my questions during a phone call from San Francisco. Since we had known each other a long time, this was almost as good as a face-to-face meeting, and I asked him to tell me how he had become involved.

Woodward told me of being woken in the early hours one morning by a phone call. It came from Norway, which explained the awkward timing, and the caller was James LeVoy Sorenson. Sorenson, who died in 2008 at the age of 86, was the richest man in Utah, and although Scott had never met him, he certainly knew who he was. Sorenson had made a fortune first in real estate and then as an inventor of medical devices, notably the first modern intravenous catheter and, more prosaically, the disposable surgical mask. He was also a Mormon, having served his obligatory mission for the Church of Jesus Christ of Latter-day Saints in New England in his early twenties. While in Norway, where his ancestors had their roots, and evidently aware of the growing application of genetics to the questions of human origins, he had come up with an idea.

"Scott," he asked, "how much would it cost to do Norway?" It became clear to the by-now-wide-awake Woodward that "doing Norway" meant DNA testing the entire population of four million people. "Tens of millions of dollars, at least," was his off-the-cuff response. There was no immediate answer from Sorenson, and Woodward imagined that the figure was way too high, much higher than Sorenson had in mind. The call ended, and Scott went back to sleep assuming he had heard the last if it. He was mistaken. When Sorenson returned to Salt Lake City he arranged a meeting with Woodward in order to come up with a plan for "doing Norway." (When you are the richest man in Utah, with a fortune estimated at $4.5 billion, tens of millions must seem like small change.) To his credit, Woodward managed to persuade Sorenson that perhaps "doing Norway" was not the best way to go and managed to morph the ambitious yet geographically limited project into something with much greater promise. Why not collect DNA from volunteers all over the world along with their family histories? Sorenson quickly agreed, and Woodward left Brigham Young to lead the project at the eponymous research institute. The aim was to collect and store a hundred thousand DNA samples, an ambition realized in 2007. There was also a commercial arm, Relative Genetics, which like Oxford Ancestors and Family Tree DNA offered a service direct to the public.

Last, there has been a numerically impressive initiative with corporate backing. The Genographic Project, a joint venture of *National Geographic* and IBM, has tested DNA from 350,000 people from all over the world since it began in 2005. The project leader, Dr. Spencer Wells, who rejoices in the enviable title of a *National Geographic* "Explorer-in-Residence," is himself a geneticist, and he and I once worked in the same research institute in Oxford, though not in the same laboratory. His expertise was in the genetics of the cellular immune system, a horrendously complicated natural defense arsenal that keeps our bodies from being overrun by pathogens. Unfortunately it can take its job a bit too seriously and turn its impressive destructive powers against our own bodies, leading to rheumatoid arthritis and other autoimmune conditions. It is also the system that causes tissue transplant rejection. However, the bewildering genetic variation that underpins the autoimmune system has been seized upon by geneticists as a more sophisticated equivalent of blood groups, and put to use in differentiating populations for the purposes of finding out where people came from. I have never particularly liked this approach to unraveling the past, for reasons I have written about in *The Seven Daughters of Eve*. It meant that Spencer and I never worked closely together in Oxford. While I cut my teeth in the coral-fringed islands of Polynesia, Spencer's chosen region was the harsh and arid steppe land of Central Asia, in particular Uzbekistan and Kyrgyzstan. By the time he became the director of the Genographic Project, Spencer had wisely abandoned the genetics of the autoimmune system and switched to the stalwarts that are mitochondrial DNA and the Y chromosome.

Soon after finding the genetic link between the Y chromosome and a handful of English surnames, including my own, and while the whole field was firmly within the "research" phase—meaning that we weren't going to get any objections from our main funders, the Wellcome Trust—an opportunity came my way to test, in depth, the usefulness of the surname/Y-chromosome association. With the Sykes study and the other names used to check that Sykes was not for some reason unique, there was no parallel genealogical research. I had basically picked the volunteers randomly. So long as they shared a surname, that was all that mattered. While I was on the lookout for a surname with a well-researched genealogy behind it, I was contacted by Chris Pomery. An interesting man in many ways, he had recently returned from a spell as a correspondent in the Berlin offices of the London *Times* as well as other assignments in Prague. He has since written two successful books on DNA and genealogy that have been very useful practical introductions for genealogists all over the world. When I first met Chris he exuded—whether deliberately or not—something of the air of an international man of mystery, which added spice to his visits to my lab in Oxford.

In between his (as we all liked to imagine) murky dealings in Eastern Europe, Chris had done a huge amount of genealogical work on his own surname. He had tracked down 825 living holders of the Pomeroy name and its variants and linked them through the records to one of fifty-one named ancestors. Some of these ancestors lived a long time ago, the oldest in the 1600s, while others were much more recent. There were also different spellings to be considered, like "Pomery," "Pomroy," and "Pummery," that might or might not be genetically related. As so often in genealogical research, there were no reliable records with which to connect the different branches, and Chris approached me to see if genetics might provide the means of doing so. When he outlined the situation to us in Oxford, his project seemed to have all the ingredients we were looking for as a practical test of the Y chromosome, and we immediately decided to go ahead.

One of the advantages of working in Oxford is that undergraduate students in the biological sciences have to do a piece of original research during their final year and write a dissertation. Each year I took on at least one of these bright youngsters, and I could see that the Pomeroy project would be ideal. It would be bound to generate results for the dissertation, it was contained rather than open-ended, and above all it was original and interesting. The student in 2000 was David Campbell, and the three of us met in the coffee room at the Institute of Molecular Medicine, the location for many a planning session, to map out the details. Chris explained that the surname Pomeroy originated from the Norman adventurer Ralph de la Pomerai, who was granted a number of English

manors by William the Conqueror as a reward for his loyalty and support during the Norman invasion of 1066. The manors were mainly in
the southwestern county of Devon, with the family seat at Berry Pomeroy castle. The castle, now a ruin, is tucked away in a steep wooded valley
a few miles outside the picturesque and historic town of Totnes. It was
sold in 1547 to Edward Seymour, the first Duke of Somerset and brother
of Jane Seymour, the third wife of King Henry VIII. Berry Pomeroy, now
uninhabited except by the ghost of the White Lady (ghosts are considered a de rigeur accessory in all the best castles), still belongs to the Somerset family, now headed by John Seymour, the nineteenth duke.

We picked one member of each of the fifty-one groups that Chris
had identified as being related through the records, and David began
the (at that time) laborious process of generating their Y-chromosome
signatures. By then we had expanded the number of markers we used
from four to seven, still very limited by today's standards. Even so, when
the results were all in, they were very revealing indeed. Despite the fact
that the name originated in a single individual, Ralph de la Pomerai, the
genetics revealed that the fifty-one volunteers belonged to at least eight
genetically unrelated branches. Chris expressed little surprise at this, but
I was certainly taken aback. In the other names that I was researching at
around the same time, the general rule had been that when a surname
was rare and its geographical distribution localized, there was usually
just one major branch descending from a single ancestor. Of course that
would not be the case in a common occupational surname like Carpenter
or Fletcher, or names that were clearly derived from a feature of the landscape such as Bush, Hill, or Greenwood. But for Pomeroy, confined as it
was to Devon and Cornwall and having a very definite origin, I thought
something else must be going on. But from a very practical point of view,
the genetics had shown Chris and his fellow Pomeroy researchers which
of the links between the fifty-one groups could be followed up in the
records with confidence and which would be a waste of time. Since then
Chris and his colleagues have expanded the Pomeroy project to become
"probably the most advanced surname project in the world," according to

the project Web site. They have thoroughly investigated the genetic links between alternative spellings of the name: "Pomery," "Pummery," "Pomroy"; in Australia, "de Pomeroy" and "Pommeroy"; and, in the United States, "Pumroy." Interestingly, men with some spellings, like "Pomery," usually share the same Y chromosome, while with others the genetics bears little or no relationship to the spelling, a reminder to genealogists everywhere that names can mutate much faster than Y chromosomes.

In some cases we know the precise origin of a surname, and can be certain that it was unique. Take Dyson as an example, which careful research by Dr. George Redmonds has shown to be an unusual case of a matronymic name meaning "the son of Di." The Di in this case was an unmarried cattle rustler called Dionissia of Linthwaite who named her son, John, born in 1316, after herself rather than his father. With this unusual single origin, it is far less surprising that the name Dyson has a very tight association with a particular Y-chromosome fingerprint, which also has the advantage of itself being quite rare in the general population. In the case of Pomeroy, where there is a named ancestor, Ralph de la Pomerai, what is the cause of this unexpected surplus of genetically different ancestors all with the same rare surname? Here we see for the first time that the rules that apply to the peasants, and I have to include the Sykes line here, do not apply to the nobility.

In feudal estates, like that granted to Ralph de la Pomerai and maintained by his descendants, there are two forces at work that can disengage the surname of the lord of the manor from his Y chromosome. There is no doubt that the privilege of the position was accompanied by increased mating possibilities. One of these was enshrined in the custom of *droit du seigneur*, when brides-to-be had to endure a night with the lord before beginning normal married life. When a son was born as a result, he would have the Y chromosome of the lord of the manor, but the surname of the woman's husband. Sometimes the son would be given the lord's surname in exchange for material support. That would not necessarily disengage the surname from the Y chromosome, unless the son was actually fathered by another man, the woman being happy to keep quiet and take the money.

The Pomeroy case also showed how genetics can reveal, or dissolve, links to other names with similar spellings. Alternative spellings of the same name abound, and pose a particular puzzle for genealogists. Sometimes surnames are deliberately changed, and there are abundant examples of European immigrants to the United States and to Britain who have anglicized their surnames. Gutmann to Goodman, Beckmann to Beckham, and so on. It is very common. Later the name may be reverted to the original. But mistakes, deliberate or otherwise, by officials who are either processing immigration papers or recording births, are a common historical cause of alternative surname spellings. In medieval Britain, when most of the population was illiterate, it was the job of the local parson to register the births in his parish. Inquiring what the surname of the baby should be, it was an easy mistake to mishear the parents and write down a slightly different spelling, which, being unable to read, the parents were not in a position to correct. Even now, whenever I am dictating my name for a form or booking a restaurant over the phone, more than half the time I become Mr. Skyes. The only place this never happened was when I was researching the Sykes heartland in West Yorkshire. There they got it right every time.

Alternative spellings are very common in the United States, either as deliberate adoptions to disguise a foreign origin or by the carelessness of clerks at Ellis Island, New York, or other entry points to the United States. Genetics has been helpful in linking American citizens with altered surnames back to their European origins. There are by now hundreds of examples, but one of the first in which I was involved, through Oxford Ancestors, was when we were approached by two American families, the Lehmans and the Bachmanns, with very similar requests. They were both looking to establish links back to presumed ancestors in Germany and Switzerland. Within each of the families, their own research in the records had uncovered an array of alternative spellings in the United States, so there were good reasons to turn to genetics to check whether the genealogical connections that had been made within each family were real or not.

To cut two very long stories short, the Y chromosomes in both families gave a very clear idea of the different branches and the range of spellings within them. Among the Lehman family, there were three clear branches defined by Y chromosome signatures, but the distribution of alternative spellings was more or less random between them. One branch contained mainly Laymans and Laymons, but another Layman clearly belonged to another branch with Lemons, Lemmons, and Lehmans while a third had a bit of everything; Layman, Laymon, Lemon, Lemmon, and even La Mance, all of them genetically related to one another. Among the Bachmann family, on the other hand, there were only two alternative spellings, Beckman and Baughman, but that was no help in defining the family structure as both were found within each of the four main branches defined by their Y-chromosome signatures. Once again genetics had shown the dangers of assuming that individuals with the alternative spelling of a surname were necessarily related. With the real branch structure now revealed in both families, they were able to link members of each branch to the correct relatives back in Europe.

In England it is surprising how many people claim to have ancestors who, like the Pomeroys, "came over" with William the Conqueror in 1066. This is a generally harmless boast, but I doubt if it is true in most cases. I hope it has not come as too much of a disappointment to the Pomeroys who have been proved to be genetically separated from Baron Ralph. I suspect, however, that the news will not have diminished their aspirations to a distinguished Norman ancestry.

The English are not alone in craving a noble ancestry, and genetics has reignited this desire by opening up the possibility of proving it. Again, an early case in which I became involved concerned the Cloughs of New England. The Clough Society of North America is one of hundreds of one-name groups that, very early on, saw the benefit of genetics in testing the links between their members. The society had the advantage of having an experienced genealogist, Sheila Andersen, who got in touch with me to test the link she had found, by working through the records, that the U.S. Cloughs were related to the Welsh nobility of the same name. Accordingly my lab tested Clough Society members and Sir William Clough, the head of the Welsh noble branch of the family. Like many American genealogy groups, the Cloughs enjoy coming to Britain to search out the locations of their ancestors, and Sheila is very good at organizing these tours. This one included the fairy-tale village of Portmerion, built by another ancestor, the late Clough Williams Ellis. The location for the denouement, where I was to reveal the results of the Y-chromosome tests, was to be in St. John's College in Oxford. Not my own college, but considered appropriately medieval for visiting Americans. This is not the only time that visiting groups, especially television documentary makers, have transplanted my perfectly good study to more "authentic" surroundings. No substitution was more impressive than when I was engaged to help unravel the Welsh ancestry of the actress Susan Sarandon. The producer decided that the appropriate location for our scene together should be none other than the stately home of the Duke of Marlborough at Blenheim Palace, a few miles north of Oxford and known the world over as the birthplace of Winston Churchill.

Even the fabulously wealthy St. John's College could not match the opulence of my "study" at Blenheim. Even so, the richly decorated college room with a fan-vaulted ceiling was a suitably historic venue, and after my introduction to the mechanics of the genetic tests, I came to the results. Most of the group of seven were women. They were Cloughs, all right; but, not having Y chromosomes themselves, had obtained the vital DNA samples from their male Clough relatives. In two cases this was from their husbands, so the women concerned did not have any strictly genetic Clough ancestry, but that did not diminish in the slightest their thirst for a touch of nobility in the family. I could sense the anticipation in the room when I flashed up the slide of the results, which was in the form of a color-coded table of the Y-chromosome fingerprints. Muffled yet audible sounds of delight filled the room as it became clear that the Clough Society Y-chromosome fingerprints matched the sample from Sir William. All except one, which was entirely unrelated. This chromosome, unfortunately, came from the holder of a high office in the society.

Sometimes the records are ambiguous and point in two different ancestral directions. This was the case with the Lockwood family in America. They had been researching their English origins for several years, but the majority had been unable to discover where their ancestor had lived. Only one had been able to follow a paper trail back to an Edmund Lockwood, born in the Suffolk village of Combs in 1574. The others did not know whether they too could claim Edmund as their ancestor or whether they were from somewhere else altogether. To try to untangle this, the family had spent many years combing the records in the two English locations where Lockwoods were concentrated. One was in Yorkshire, in northern England, while the other was in Suffolk, in East Anglia. But so far the effort had been in vain. Uncertainty always saps enthusiasm, so the Lockwoods had a very good reason to want to know which of these two locations was home to their ancestor.

This was a case ripe for genetics, and before long we had enrolled half a dozen male Lockwoods from either the old weaving town of Halifax, which our surname map soon showed to be the present-day epicenter of the Yorkshire branch, or from around Ipswich, the county town of Suffolk. Their Y chromosomes immediately revealed that the Yorkshire and Suffolk Lockwoods were unrelated to each other, but that within each region their chromosomes matched. When six American Lockwood chromosomes were tested, they were all identical, and not only did they match one another, they also matched the Suffolk Lockwoods but not their Yorkshire namesakes. At a stroke the ambiguity had been eliminated, and the American Lockwoods could concentrate their considerable energies on Suffolk, freed from the gnawing anxiety that they were wasting their time. Which, in Yorkshire, it turned out they had been.

The link I had discovered, almost by accident, between surnames and

Y chromosomes has certainly found a use in testing the genealogical links between men with the same surname and illuminating the process of alternative spellings, as we have seen. This is all very interesting in what it says about surnames. But it also says a lot about men, and seemed to me to have the potential to explore the contrast in mating habits between the aristocracy and the peasantry, which my work on the genetic history of Britain and Ireland for *Saxons, Vikings, and Celts* had indicated was extreme. In the Pomeroy case there was circumstantial genetic evidence that the surname of the original Norman baron had been adopted by other men in the vicinity, and I have suggested how the behavior of the feudal lord and his descendants may have played a part in this. However, the Pomeroy case did not prove that the baron's Y chromosome had been dispersed within the local population, as I suspect it very well might have been.

A few years later I had the opportunity to carry out a direct test of this phenomenon, not around Berry Pomeroy, but on another large estate, this time in Wiltshire, in southern England. The estate was Longleat, and I had gotten to know the owner, Lord Bath, when he had asked me to see if he was related to "Cheddar Man," a nine-thousand-year-old fossil excavated from Cheddar Caves, on his estate. I had just published my DNA results from Cheddar Man, which showed that this ancient relic was related, through his mitochondrial DNA, to a history teacher in the local school. The story encouraged much mirth, especially in the United States, as it reinforced the bucolic, stay-at-home image of the English in that the descendants of Cheddar Man had taken nine thousand years years to travel three hundred yards down the road. It turned out that Lord Bath was not related to Cheddar Man, but his butler, Cuthbert, was. If I expand on this amusing anecdote much longer I will be guilty of repeating myself, as I described the episode in *The Seven Daughters of Eve*.

However, its relevance here is that when I was looking around for a test for what I came to think of as "aristocratic diffusion," Longleat fitted the bill very well. The estate had been continuously in the hands of the Thynne family since the sixteenth century. More to the point, the majority of the estate workers, perhaps the most likely vectors of aristocratic diffusion, had come from the estate village of Horningsham a mile or so away. Lord Bath himself had become very enthusiastic about genetics and welcomed my inquiry. When I went to visit him in his penthouse at Longleat I was very pleased to see that he had mounted his mitochondrial family tree from the Cheddar Man case on the wall near his enormous chestnut desk. Lord Bath is famous as a confirmed polygamist, having at least fifty "wifelets" (his own description), who are rewarded with a cottage on the estate after sufficient years of service. Other than sexual pleasure, one objective of polygamy is reproduction, but, given that his offspring are still in single figures, I was not sure whether this has been as great a success as Lord Bath would have liked. However, the prospect of discovering that his ancestors had sired the population of Horningsham clearly had a certain appeal, and Lord Bath gave his permission for the project to go ahead. In fact he did a lot more than that, as we shall see.

As I was going home I stopped off at Horningsham parish church to have a look round the churchyard. It was November and getting dark, but I was still able, with the help of my flashlight, to make out the names on some of the crumbling headstones. I was looking for surnames, and sure enough, there were only a few—Trollope, Long, Carpenter—with several examples of each. This is what I had expected and hoped for. It is a sign of a stable and static population, where surnames are winnowed out over the generations as families have no sons to carry them on. Genealogists know this phenomenon very well as a surname becomes less and less common and then disappears. Rather cruelly in my opinion, the surname is said to have "daughtered out." In the United States, Dearborn is a numerous and well-known name. Almost all of the 5,000–10,000 U.S. Dearborns are descendants of one man, Godfrey Dearebarne, who arrived in New England in 1639. But while Dearborns have thrived in the United States, the name Dearebarne has disappeared from England, where it originated, the hapless victim of "daughtering out."

In the churchyard at Horningsham, daughtering out had reduced the variety of surnames to just a handful. This is very common in rural England, which I remember from my time as a postman in university Christmas vacations at my parents' home in a village called Dedham on the Essex-Suffolk border. Being deep in the country, there were no house numbers. And there seemed to be only two surnames: Ablitt and Matthews. It was very frustrating trying to find the correct destinations for my consignment of Christmas cards, another unintended consequence of "daughtering out." However, in Horningsham it worked to my advantage, and I set about contacting all the current residents with the few surviving surnames from the churchyard. Three months later, helped by the indomitable matriarch of the village, Vera Trollope, I had enough volunteers for the project to proceed.

Longleat is one of the most beautiful stately homes in the whole of Britain. Built during the reign of Elizabeth I in 1580, it has a symmetrical elegance not always found in houses of that period. Its position in a shallow bowl of land embraced by low rolling hills means that the first sight you have of the house is a distant one, looking down from the wooded rim set in parkland laid out long ago by the prince of landscape gardeners, Capability Brown. Longleat was built by Sir John Thynne, a man who was not born into great wealth but who worked his way up by administering the affairs of others, finally becoming the right-hand man, or steward, to Edward Seymour, Duke of Somerset, whom we have already met as the owner of Berry Pomeroy castle. At the same time Thynne began to build up his own property holdings, culminating in the purchase of Longleat in 1540. In 1549 he married into the wealthy Gresham family and began to plan the present house on the proceeds of his wife's not inconsiderable dowry. Later the same year Edward Seymour lost not only his power base but also his head, but Thynne avoided execution and was imprisoned in the Tower of London. He was eventually released and, warned off politics, settled into country life at Longleat, where he and his wife, Christian, had nine children, with a further five, all sons, coming from a second marriage following Christian's death. The Thynne dynasty continued to thrive in succeeding generations as their titles reflected their gradual ascent through the ranks of the peerage, first as baronets (1641), then viscounts (1682), and finally marquesses in 1789. Alexander Thynn, the current Lord Bath, who dropped the final *e* from his surname during the 1980s, is the seventh marquess, and a direct patrilineal descendant of Sir John. Thus he has inherited Sir John's Y chromosome. The question was, could I also find the Thynne Y chromosome among the good people of Horningsham?

The day of the DNA collection had arrived, a brilliantly sunny Saturday in late February, and about fifty villagers had assembled in Longleat's sumptuous Red Library. I arrived with two assistants: my son Richard who, at fifteen, was already a veteran of several sampling expeditions all over the world; and Charlotte, a Danish graduate student from Oxford who had answered my advertisement for research assistants. Before the sampling session began we had been invited to a drawing room to meet Lord Bath and his weekend guests. I remember feeling slightly nervous going into the room and being met by the eyes of a dozen or so people arranged on comfortable sofas and easy chairs. Lord Bath was there, of course, wearing a trademark embroidered skullcap, and he introduced us as the DNA collectors. Curled on the floor at his feet lay a cream Labrador. By his lordship's side a strikingly elegant woman, her long blond hair interlaced with colored beads, reclined on a green velvet chaise longue. In an attempt to defuse my apprehension with bravado, I walked straight up to her and said, "We'll start with you." That is how I first met Ulla, my future wife.

The magnificent Red Library, with its leather-bound volumes filling shelf upon shelf of gilded oak, had been laid out with chairs that were filled with eager Horningsham residents. After a short speech by Lord Bath introducing the purpose of the day, Richard, Charlotte, and I set about collecting the DNA by means of cotton swabs rubbed on the inside of the cheek. I used to do this myself until the occasion, in the Shetland Isles, when an elderly lady's false teeth came loose and clamped the swab fast. After that I asked people to do it for themselves. Before long everyone had given a sample and their consent. Even the dog, whose name was Boudicca, joined in, though I was obliged to sign the consent form on her behalf. This had been a very special day.

Returning to the scientific purpose of the visit, I was looking to match Lord Bath's Y chromosome with men from Horningsham. If I found a match it would not necessarily mean that Lord Bath himself was the father of the man, though that was a formal possibility, but that the Thynne Y chromosome had escaped from Longleat House through the energies of one of his ancestors. The prospect appealed to Alexander, which is why he not only gave permission for the experiment but was wonderfully generous in making the arrangements, not only for the original collection but for the day when I returned to Longleat later that year to announce the results.

This meeting took place in the vaulted undercroft, where chairs and tables had been arranged around the stout pillars that supported the house. I left the announcement until lunch was over. First I went through the results for each family that had taken part. Within each surname the Y-chromosome signatures had been the same. I wasn't surprised by this, so was taken aback when one of the Trollopes asked, "Are you saying that we are related to the Trollopes from past the church?" "Well, yes," I replied. "No, that can't be right, we are from completely different families," came the response. I had clearly scratched a deep sore in the village.

Next I presented the results of the DNA tests from the other families, and, like the Trollopes, they were identical within each one. But did any share the distinctive Thynne chromosome? The time had come. It felt almost like an Oscar moment when the Academy Awards are announced. "And the winner is . . . George Long." George Long, a gentleman about fifty years old, rose from his chair and came up to be photographed with his now-relative, Alexander, Lord Bath. Both men were descended from the same man, one of Lord Bath's ancestors, although no one knows which one. Speaking entirely formally it could have been the other way around, but to me this was proof of aristocratic diffusion, from Longleat to Horningsham. But it was not overwhelming. The other families in the village did not have the Thynne Y chromosome, so had not sprung from the loins of Longleat. Attempting to compensate for any disappointment on both sides, I reminded the audience that, powerful though it no doubt is, genetics cannot detect unfruitful ancestral seductions.

7

The World's Biggest
Surname Project

n the summer of 2009 a remarkable event occurred in Edinburgh, the capital of Scotland. On the greensward overlooked by the volcanic outcrop of Arthur's Seat and next to Queen Elizabeth's official Scottish residence, the Palace of Holyroodhouse, the greatest gathering of the clans for more than two hundred years sprang into life. The gathering was the centerpiece of "Homecoming Scotland," an event that ran from Burns Night, the two-hundredth anniversary of the birth of Scotland's most celebrated bard, Robert Burns, on January 25, to Saint Andrew's Day, the national saint's day, on November 30. Genealogy and family history was one of the five major themes of the homecoming, which, as the name suggests, was designed to attract people of Scottish ancestry to visit the land of their ancestors. "For every single Scot, there are thought to be at least five more overseas who can claim a Scottish ancestry," ran the blurb.

The gathering attracted 47,000 people from all over the world, a large proportion coming from the United States. In the tented "Clan

Village" that spread across Holyrood Park were representatives of 125 Scottish clans, whose chiefs had convened the previous day in the nearby Scottish Parliament to debate the role of the clans in today's world. On the evening of July 25 some twenty thousand people lined the Royal Mile to watch a parade of eight thousand clan members with their pipe bands march up the ancient cobbled street from Holyrood-house to Edinburgh Castle. It was a magnificent and stirring spectacle at the end of a brilliantly sunny day, one I will never forget. The Duke of Rothesay, the title that Prince Charles uses when he is in Scotland, reminded the throng when he opened the gathering that clan life had quieted down over recent years. "Thankfully, in 2009, the lives of clan chiefs and their clansmen, both in Scotland and abroad, are somewhat less blood-soaked and unhappy than those experienced by thousands of their ancestors."

Out of Oxford term time, I spend a lot of my year in Scotland, drawn by the wild beauty I first encountered when researching *Saxons, Vikings, and Celts*, and it seems to me that a lot of what happened in Scotland is relevant to this portrait of the United States. First from a practical point of view, hundreds of thousands of Scots emigrated to the New World and so have a solid genealogical connection with Scotland. This is fused with the strong emotional bond that motivates tens of thousands of Americans to research their Scottish ancestors and compels many to visit.

Jane, from Texas, posted this comment on the gathering: "I was an American visitor with Scottish heritage who visited Scotland for the first time. I am now and forever 'in love' with Scotland and I can hardly wait to return 'home' and see more of your lovely country." Of course it's easy to post a comment on the Web, but Jane's comments are both sincere and to the point. They are repeated over and over again in remarkably similar fashion, not just by émigré Scots but by people of all nations who have come to live in America. Like German and Russian Jews; like impoverished families from Ireland, Sweden, and Norway; like Chinese,

Japanese, and all the multitudes from other countries that flooded into America in the nineteenth century, the Scots share a devotion for America but remain deeply rooted to their homeland. "America first, Scotland always," is how one articulate farmer from Wisconsin summed it up. For the most part descendants of these European immigrants, in common with their counterparts from Asia, present no great general mysteries that require genetics for the unraveling. Russians came from Russia, the Irish came from Ireland. But where genetics has had an impact is in individual quests, solidifying the arcane, almost mystical sense of connection that people feel with their ancestors.

In Scotland the scaffolding for this connection is articulated in the clan system, which has its parallels in every society at some point in its evolution. As the Duke of Rothesay hinted in his speech to the assembled clansmen and women at the gathering, the history of Scottish clans is one of almost unrelenting and bloody wars between rivals. The clan was the unit of allegiance and protection, just as it was for American Indian tribes. And just like them, Scottish clans raided their neighbor's land for cattle, treasure, and slaves. Just like Indian or African tribes, too, membership rules were flexible depending on the circumstances, and this is evident from the genetics, as we shall see. Scottish clans, like tribes, infuse their members with the fierce loyalty once required in battle, now fought out in the gentler climate of tartan design and the issue of coats of arms by the Court of the Lord Lyon, Scotland's chief herald. In my experience the fervor is even greater among the descendants of Scots who settled in America when compared with their counterparts who stayed behind. At the gathering it was the members of the American branches of the clan societies who dressed with that little extra attention to detail. Just as the Amana colonies in Iowa have preserved their German language, dress, and culture with an enthusiasm rarely encountered in Germany itself, Scots Americans perpetuate the attachment to their roots in all manner of ways—in dress, song, and even in cooking. It is said that more haggis— the Scottish national dish of minced sheep lungs and oats wrapped in intestines—is sold in America on Burns Night than in the whole of Britain. (Haggis tastes a lot better than it sounds.)

I first became involved in testing the genetics of Scottish clans at the tail end of my work on the surname/Y-chromosome association. I was aware that clan surnames were different from their English counterparts in that sharing a clan name did not necessarily imply a direct relationship to the clan chief. Indeed, there are so relatively few Scottish names that a tight genetic relationship between clan membership and the Y chromosome would have implied an impressive breeding effort by only a few men over several centuries. While by no means denigrating the invigorating properties of porridge or whisky, and the opportunities for easy copulation afforded by the kilt, I had never imagined anything significant would come out of a study on Scottish surnames—such was the high level of name adoption that I had assumed. I am glad to say I was completely wrong, but my lack of enthusiasm for investigating a link meant that I did not discover my error of judgment for two years.

I was eventually set on the right track when my research team and I set out to compile our genetic map of Britain, which went on to provide the material for *Saxons, Vikings, and Celts.* For reasons to do with the way our research budget was structured, we first set out to cover Scotland before moving on to England and Wales. Taking samples from volunteers at blood donation sessions throughout Scotland, we collected DNA from several thousand volunteers over the course of two busy years. We had their written consent, and thus their names and addresses, but originally only as a means of circulating the project results. At the end of the two years my colleagues and I sat down to analyze the enormous set of results. This is something I really like to do. It sounds terrifically dull, but seeing and then interpreting raw lab data in the form of squiggles on a sequencer read-out gives me a quiet thrill even now. Like my colleagues, I had developed a nose for the unusual. I had noticed one particular Y chromosome that belonged to the clan of Sigurd, named after the Norse god. This was my own romanticized nomenclature, sadly reduced in my

opinion to the far-more-prosaic-sounding "haplogroup R1a" by almost
everyone else. This chromosome, of Norse Viking origin, was not common in Scotland. It was not confined to any particular region, but was, if
anything, more frequently encountered in the Highlands and the Hebrides than anywhere else. Only when I got around to looking at the names
of the volunteers did I feel even a flicker of interest. Men with the three
surnames MacDonald, MacDougall, and MacAlistair were the only ones
to have this particular chromosome. It was when my graduate student
Jayne Nicholson pointed out that all three names were supposed to be
linked to a common ancestor that I began to get excited.

This ancestor was Somhairle mac Gillebride, better known as Somerled,
a twelfth-century Celtic hero who had rid the western seaboard of
Norsemen. According to the traditional genealogies, all three of the
clans were linked to Somerled through his sons and subsequent generations of patrilinear descendants. Over the next few weeks Jayne contacted dozens of men with these names while I concentrated on the five
living clan chiefs. To cut another long story short—I have written about
this at greater length in Saxons, Vikings, and Celts—all five chiefs shared
the same rare Y-chromosome signature, as did a substantial proportion of our surname volunteers. This was a testament to the fidelity of
the clan chiefs' wives over the last nine hundred years, exceeding even
the devotion of the Mrs. Sykeses. I did not discover a single nonpaternity event even though the clan chiefs' ancestral lineages were made up
of eighty seven independent father-son generations. But although the
impeccable behavior required to maintain the correspondence of name
and chromosome over such a long period might have been true for their
wives, the chiefs had evidently not contained themselves quite so well.
A high proportion of the men with the three surnames also carried the
chiefly Y chromosome, even though none realized they were related to
the clan chiefs and therefore also descended from Somerled. This was
not so much aristocratic diffusion as saturation.

About a third of male MacDonalds in the study had inherited

Somerled's Y chromosome, while in the other clans the proportions were even greater. Among the MacDougalls 40 percent carried the chiefly Y chromosome, and among the MacAlastairs almost half were Somerled's patrilineal descendants. I was puzzled by the clear difference in these ratios until I discovered that, historically, the clan fortunes had been very different. Clan Donald is, and has been for centuries, the largest and most powerful of the three clans. Clan Dougall, descended from Somerled's eldest son, had once been much more influential but had most of their estates confiscated after attaching themselves to the losing side in the war that put Robert the Bruce on the Scottish throne at the beginning of the fourteenth century. Clan Alastair, sadly, had never been of much consequence. The proportion of men with the surname who also carried the Somerled chromosome was inversely proportional to the size and power of the clan. Clan Donald was the strongest, yet the proportion of genetically related male MacDonalds was lower than in the other two clans. Even so, at 30 percent it meant that an astonishingly large number of men carried the ancestral-clan Y chromosome. I had certainly been wrong to assume that there would be only a very weak connection between surnames and genetics among the Scots.

The inverse ranking among the three clans suggests to me that straightforward name adoption was higher in Clan Donald than in the other two. Name adoption was a widespread custom, as men were encouraged to take on the chief's surname if they found themselves living on his land or fighting in his army. There was more scope for both of these in Clan Donald than in the others. Even so, large numbers of men were genetically related to the clan chiefs and had inherited Somerled's Y chromosome, and their surname, from one of their ancestors. While some may have done so as descendants of one of the chiefs' brothers, Maggie MacDonald, archivist for Clan Donald, conceded that the chiefs had engaged in their own version of *droit du seigneur*, often imparting their surname as well as their seed.

The genetics had shown that the clan genealogies were surprisingly accurate and that the ancestors of the five living clan chiefs had chosen remarkably well-behaved wives. But they had also discovered a telling example of medieval public relations. The clan genealogies emphasize Somerled's Celtic credentials by tracing his own ancestors back in time to a long line of Irish kings. However, his own chromosome is so typically Norse that I can only believe the invented Celtic ancestry was a deliberate ruse, by Somerled or one of his ancestors, to bolster the claim to the chiefship of the Gaeltacht, or indeed to the crown of Scotland. It didn't succeed: He was killed at the Battle of Renfrew in 1165 as he launched an unsuccessful invasion of the Scottish mainland. Nevertheless Somerled's ancient Irish lineage, back to such semimythical predecessors as Colla Uais and Conn of the Hundred Battles, was a cherished attribute for the acknowledged headship of the Gaeltacht and the Lords of the Isles, as later chiefs of Clan Donald were to become.

As my research into Clan Donald was drawing to a close, an American clan member, Mark Macdonald, a Texas lawyer from Dallas, had become aware of the potential of genetics. Like Bennett Greenspan in Houston, Mark had read the *Wall Street Journal* article on the Cohanim and, like Greenspan, had begun to wonder what genetics might do for his own historical clan research. He got to hear about my earlier work on Clan Donald and the identification of Somerled's Y-chromosome signature, shared not only by the five living clan chiefs but also by a substantial proportion of clan members. He then set about organizing the Clan Donald DNA Project, since when it has grown to become the largest genetic genealogy project in the world based on a single family group. It is coordinated from Illinois by Professor J. Douglas McDonald, a professor of physical chemistry at the Urbana-Champaign campus of the University of Illinois. By February 2011, one thousand members of the clan had joined the project and sent their DNA for testing.[1] The Clan Donald project serves as a fine example of the way in which DNA has helped to map the structure of a large group of men with something in common, in this case their membership of a clan. It also acts as an example of how a relatively small-scale academic study has been adopted and amplified not by scientists like me but by the individuals most concerned with the outcome.

Apart from its importance as a prime example of public participation in what is, after all, a heavily science-based project, a major question is to what extent the DNA results have influenced the historical account of the clan history. The clan has always had excellent historians who have had access to the extensive archive of the Clan Donald Centre at Armadale on Skye, not far from where I am writing this section. From the 1400s to the present day, or at least until genetics raised its head, the firm consensus among Clan Donald historians was that Somerled was descended from the Irish king Colla Uais, who lived around AD 330 and whose kingdom was centered on the northwestern province of Ulster. Ancient Irish histories linked Colla Uais to another important Irish king, Niall of the Nine Hostages. Niall lived around AD 400, around the time that the Romans abandoned Britain to save Rome, leaving the west coast vulnerable to raids from Ireland. Niall, as his sobriquet suggests, made his living by capturing and then ransoming. His most famous captive was Saint Patrick, who went on to become the patron saint of Ireland. Both Niall and Colla Uais were, according to the body of Irish history, descended from Conn of the Hundred Battles, who was High King of Ireland around AD 175. The accuracy of Irish historical accounts suffers from a lack of written evidence, largely because, unlike most of Britain, Ireland was never occupied by the Romans. The Irish language, Gaelic, was not written down at the time, so everything was passed down by word of mouth.

Even written records can never be entirely relied upon, however, as authors were dependent on sponsors who were as much, or more, interested in creating a favorable impression than in historical accuracy— so much so that some modern Irish historians question whether Conn of the Hundred Battles was a real person at all or a mythical sun god. Others have questioned whether the common descent of Colla Uais and Niall of the Nine Hostages was a later invention of the O'Neills to legitimize their invasion of Ulster. Inventing claims like this was widespread in medieval Britain, and none was more brazen than that of Edward I who deliberately linked himself to the legendary King Arthur to back up his attempted invasion of Scotland, which he justified as an honest attempt to reunite Arthur's fragmented kingdom.

Unlike King Arthur, we can be quite sure that Niall of the Nine Hostages did exist, and that he was the founder of the eponymous Ui Neill clan. A team at Trinity College, Dublin, led by Professor Dan Bradley, compared the Y-chromosome fingerprints of Irishmen with surnames linked to the clan, principally O'Neill, and found that they often had a chromosome in common, indicative of shared patrilineal descent.[2] The Ui Neill chromosome is extremely common in Ulster today, with 27 percent of men having inherited it. Not all of these were O'Neills or had surnames with a genealogical connection to the Ui Neill, so it looks to me like another example of aristocratic leakage, Irish style. The Ui Neill dominance in Ireland, and Ulster in particular, squeezed the other Irish clans and many left to establish Gaelic settlements in western Scotland around AD 500. There followed a seesaw conflict with the original Celtic inhabitants, the Picts, which only came to an end when Scotland was united under its first king, Kenneth mac Alpin, a Gael, in AD 843, as a response to the threat from Norse Viking raiders.

Clan Donald genealogists connect Somerled to Colla Uais through a direct patrilineal line with twelve, or thirteen in some sources, named ancestors. There is an obvious discrepancy here, as the average generation time would have been almost sixty years each to stretch from AD 330 to AD 1100, the approximate year of Somerled's birth. Although some of them doubtless took advantage of their elevated status to continue having children into old age, the first son would surely have been born well before his father reached sixty. While this is a fine Gaelic pedigree, it must be mistaken, because Somerled's Y chromosome has all the hallmarks of a Norse Viking. The clan of Sigurd is decidedly Norse. Chromosomes from his clan are very rare in Britain in those regions not settled by Vikings, such as inland Wales and Ireland. The only places in Britain where this Y-chromosome clan is found in abundance are

Orkney and Shetland with their well-known history of Norse settlement. Indeed, these two archipelagoes belonged to the king of Norway until 1468, when they were annexed by Scotland to make up for a dowry that was never paid. Y chromosomes in the clan of Sigurd are very frequent in Scandinavia, especially Norway, and also abundant in Iceland, which was settled from Norway from the tenth century. Furthermore one particular marker is able to distinguish Sigurd chromosomes from Norway from those with an origin in Denmark, whose Viking fleets raided the east coasts of England and Scotland. The Somerled chromosome belongs firmly to the Norwegian type. Consequently there is very little chance of a Sigurd chromosome, like Somerled's, being found among Irish Gaels in the fourth century, the time of Colla Uais.

How can these two different conclusions, the historical and the genetic, possibly be reconciled? The Clan Donald Society explanation is that, at some point in the line of Irish kings that make up the well-researched Gaelic pedigree from Colla Uais to Somerled, there was what boils down to a nonpaternity event. At least one man was not the biological father of the son recorded in the traditional genealogy. Mark Macdonald has suggested a colorful explanation that points a finger to the link between Fergus and his son Godfrey, born around AD 830. This was a turbulent time in northern Europe, and all countries bordered by the sea lived in fear of a Viking raid. The Vikings took full advantage of their reputation and developed a very successful invasion strategy with all the features of a protection racket.

The best known of these adventures was planned and executed by Hrolf, better known as Rollo, from Ålesund on the west coast of Norway. Camping out at the mouth of the river Seine in northern France, he blockaded Paris, which lies a hundred miles upstream, and sent regular raiding parties to wreak havoc in the capital. After a while the French king, forever afterward known as Charles the Simple, sued for peace and Rollo struck a deal: We will leave you alone, indeed we will promise to protect you from other Viking raids and all your other enemies into the bargain. In exchange, you will give us Normandy. Charles agreed, and, in

911, Rollo, true to his word, cleared Normandy of the king's enemies and became the first duke. One hundred fifty years later his descendant William, Duke of Normandy, led the successful invasion of Britain in 1066.

According to Mark Macdonald's ingenious theory to accommodate Somerled's Norse Y chromosome with his traditional Gaelic genealogy, Rollo wasn't the only one in the protection business. This time the Norse racketeer was Gutfrith, who tried his luck with Fergus, chief of the Mac-Uais clan from Oriel, around the shores of Loch Foyle in the north of Ireland, who were having a hard time resisting the expansion of their neighbors to the west, the Ui Neill. This time the price for protection was marriage to Fergus's daughter and a change of name to Godfrey mac Fergus to make his accession to the chiefship of the clan appear legitimate. Thereafter the chiefs carried Godfrey's Viking Y chromosome all the way down to Somerled himself.

This is history massaged for political purposes, as history so often was. It suited Godfrey to side with the Gaels, and he used his credentials to gain lands in Scotland by helping Kenneth mac Alpin finally to defeat the Picts and become Scotland's first king in AD 843. Godfrey died ten years later, in 853, by which time he had added the title Righ Innse Gall, "king of the foreign lands," to taoiseach of Oriel. The Gaelic ancestry concocted by Godfrey was used to good effect by Somerled and his descendants, who make up the Clan Donald. Somerled himself gained his reputation for clearing the Vikings out of the west coast of Scotland and the Hebrides, so a solid Gaelic ancestry was a distinct advantage, even if he knew it to be false. Though there were hints of a Norse association in the clan, and Somerled married a daughter of the king of Norway, for the intervening 850 years his Gaelic pedigree was never really in doubt. Only in the early years of the twenty-first century did genetics finally uncover the truth. Although as yet unsupported by historical evidence, Mark Macdonald's theory is well worth serious attention and certainly provides future Clan Donald historians with a target for their researches. If he is right, and Somerled's Norse Y chromosome entered the clan genealogy with Godfrey mac Fergus in the ninth century, then how appropriate it is that the current chief of the clan, Lord Macdonald, one of the volunteers in my original DNA study and who does carry the Norse Y chromosome, is also Godfrey. All this from a withered fragment of DNA that travels blindly from father to son, oblivious to the bloodshed and betrayals that smooth its path from one generation to the next.

Because it is the largest family study so far, the Clan Donald DNA project has been able to examine the internal intricacies of the clan structure. As such it is a superb example of the potential for genetics to place individuals on particular branches defined by DNA, and to examine in greater detail how this compares with what is already known from the careful examination of the genealogical records that has been done, first in parallel and then in combination.

I have seen close up the way in which genetics, and geneticists, sometimes proclaim that their science makes all that went before both unnecessary and irrelevant. I had the good fortune to experience at firsthand the first forays of modern genetics into medicine and, through ancient DNA, into archaeology and both followed rather similar patterns. It must be disconcerting to have spent a lifetime in an area, to have researched a particular subject in great depth, only to be told in no uncertain terms by a young geneticist (which we all were once) with no experience in your specialty that you were completely mistaken, that your expertise was eclipsed by genetics and had no future. I can remember the brashest "young Turks" in the field saying things like "biochemistry is dead"—a relief to many, I am sure—but meaning that there was no longer any need to research things like enzymes and proteins. All you needed to know was the DNA sequence and everything would flow from that. This was pure hubris, and there will be many a medical specialist, now retired, who grins with quiet satisfaction when he reads that, ten years and billions of dollars later, the Human Genome Project has achieved very little as far as alleviating or even untangling the suffering caused by disease. The claims of ten years ago that we were about to witness the greatest medical advance since antibiotics have proved, thus far, to be thoroughly hollow.

However, in genealogy the two approaches have not clashed. This is partly because few scientists strayed into the field, so it never became a great source of career advancement or research funds, and also because the genealogy world was so well organized and mature. Genetics could never on its own replace the painstaking effort in the records, but it can and has added a brilliant new tool to the genealogists' kit. The two do work hand in hand very well, and the Clan Donald Project, orchestrated from the United States, is a prime example.

Although it might be thought that the official clan historians might resent the genetic conclusion that Somerled and the chiefly line of Clan Donald has a Norse rather than Gaelic origin, that has not happened. Genetics is not a dirty word within the clan, as the sheer size of the DNA project testifies. So, in addition to the discovery of the chiefs' Norse ancestry, what else has been revealed about the clan? I am well aware that in genealogy there is nothing as interesting as your own family history, and nothing as dull as someone else's, so I hope you will forgive my delving deeper into this clan's genetic history by way of illustration. Every family is different, of course, but most of what happens in other families also happens in Clan Donald.

Let us start with the chiefly line, which is Somerled's own. We can be sure of that because the same Y chromosome is found among McDougalls who descend from Dugall, the elder brother of Ranald, whose patrilineal descendants are the clan chiefs of Clan Donald. The number of Y-chromosome markers has been extended from the original seven that I used to study the clan in the late 1990s to thirty-seven and, in some cases, even more. Using the chiefly DNA as reference point, 120 of the 935 participants who had enrolled when I examined the results carry either an identical or very close Y chromosome as defined by the markers. However, they are not all called MacDonald or McDonald. There are a lot of McConnells and McDonnells and McDaniels, which confirms these as alternative spellings to Mac-Donald. There are also McKeans, a couple of Alexanders and Wilsons, a McCain, a Gordon, a Douglas, and even a Campbell, the sworn enemies of

Clan Donald since the Massacre of Glencoe in February 1692. Thirty-eight members of Clan Donald were murdered by the Campbells, who had been on the opposite side during the Jacobite uprising of 1689. What drenched the incident in infamy was that the Campbells had accepted the hospitality of the Macdonalds and had been their guests when the killing started. The massacre is still remembered in the Highlands. The memory caused an uproar only a few years ago when a Campbell was employed as the manager of the Glencoe Visitor Center, so much so that he was forced to quit. Some of these names are known spelling variants of a clan surname, such as Alexander, which is a common form of MacAlastair. Others with no obvious surname relationship, like Wilson and Campbell, may well be the result of "aristocratic diffusion," as they are likely to have come about from illicit liaisons. Somewhere in the past they got the Clan Donald chromosome without the surname. In my own study I found a couple of MacArthurs with the Somerled chromosome. The MacArthurs were hereditary pipers to the Clan Donald chiefs, which might have brought them within breeding range. They also hail from Islay, which was a Clan Donald stronghold during the fourteenth to sixteenth centuries. Either way some MacArthurs certainly carry the Somerled chromosome.

The much larger Clan Donald DNA Project has been able to look more closely at the internal genealogy of Somerled's descendants within the clan. This is a great achievement, and I think looking in a little more detail will be rewarding, not so much for what it says about an important Scottish clan, but as a superb example of what can be achieved with the combination of genetics and genealogy. Using an array of Y-chromosome markers, the Clan Donald DNA Project has compared the genetic signatures of nine men who have very solid paper genealogies back to the key ancestor, John, Lord of the Isles, who died in 1386. He is labeled "E" on Figure 2. He was the great-grandson of Donald MacRanald (C), from whom the clan takes its name. Donald was Somerled's grandson and one of his sons, Alastair (D), is the ancestor of the MacAlastairs, who, being descended from Donald, are embraced by the clan. Unlike the MacAlastairs, the MacDougalls, not being descended from Donald but linked

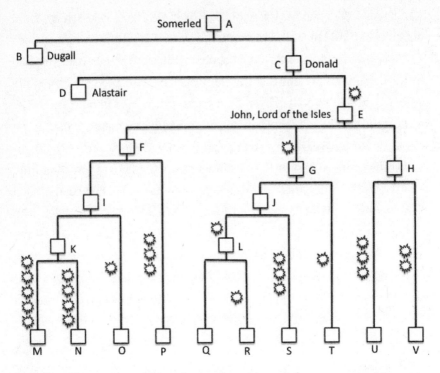

Figure 2. Genetic genealogy of Clan Donald. Y-chromosome mutations are shown as explosions (✻) adjacent to the links on which they occur. Lengths of vertical links are approximately proportional to the number of generations. Details of the mutations are given in the appendix. Labeled individuals are identified in the appendix.

to Somerled through his eldest son, Dugall (B), are not part of Clan Donald even though they share the same Y chromosome.

The men in the bottom rank (M-V) have well-documented genealogies back to John, Lord of the Isles. Three are chiefs—Ranald Alexander MacDonald, twenty-fourth of Clanranald (M); Ranald MacDonell of Glengarry (P); and Sir Iain MacDonald of Sleat (R). There are also two chieftains; Allan MacDonald of Vallay (Q) and David MacDonald of Castle Camus (S), while the other five men are neither chiefs nor chieftains. That is the genealogy, now what about the genetics? This is where it gets complicated. What I want to do now is to look at how closely a

genealogy based solely on the genetics resembles the Clan Donald genealogy as constructed from the records. It is a question I am often asked
by families who are considering embarking on a genetics project. Rather
than interrupt the narrative flow, I have put the details in the appendix.

Of the thirty-eight markers tested in this genealogy, thirteen showed
genetic variation within the clan. Mostly there was only one mutation
from the original Somerled Y chromosome at each of the thirteen markers, but one of them has mutated three times and another five times.
These recurrent mutations are a real problem in trying to reconstruct
an accurate family tree based on the genetics in the absence of a well-
documented genealogy. In this case the very same mutation has happened three separate times in different parts of the family tree. We call
these "parallel" mutations. Without the genealogy, it would look at first
sight as if there had only been one mutation, and the inclination would
be to group all the men whose Y chromosome carried it as if they were
descended from the man in whom the mutation had first occurred. This
is a familiar issue for geneticists who are trying to reconstruct past events
using genetic evidence from people alive today. My colleagues and I had
to struggle with it when dealing with hundreds of mitochondrial DNA
sequences, where parallel mutations are also an occupational hazard.

One solution is to consider all the possible family trees that could be
drawn and then work out which one requires the least number of parallel
mutations. That is a reasonable procedure for the construction of a really
large-scale tree, like the maternal genealogies of Europe for example,
when the aim is to get close enough to the true tree to be able to draw
some general conclusions. But it isn't appropriate in a family genealogy
project, where it is very important for individuals to be accurately placed.
The path of least resistance, or "maximum parsimony" as it is known in
the trade, is not designed to be absolutely accurate, just roughly right. In
fact I doubt whether the most parsimonious tree is ever absolutely accurate, and it can be misleading. Take the case in Clan Donald, where the
same mutation has happened three times. Maximum parsimony without

the accompanying genealogy would come up with a tree with the least parallel mutations, and it would be wrong.

Another important message from the Clan Donald DNA Project concerns the estimates made when two men try to find out how long ago their common ancestor lived by comparing their Y-chromosome results. It stands to reason that the more generations have passed since that ancestor was alive, the more likely that their Y chromosomes will differ from one another. This is because mutations accumulate over time, and the longer the time, the more mutations will have occurred. While this is a perfectly accurate general statement, it can be wildly misleading in individual cases. It works well when large groups are involved, so in Clan Donald, the Y chromosomes of the ten descendants of John, Lord of the Isles (E on the genealogy in Figure 2), have experienced a total of twenty-five mutations. How many generations will have passed to give that figure? The Y-chromosome-marker mutation rate is generally reckoned to be about 0.31 percent (i.e. 0.0031) per marker per generation, so the average number of generations needed to generate twenty-five mutations in thirty-eight markers at this rate is $25/(38 \times 0.0031)$, which comes to 212. Since we know the genealogy, we know the true figure to be 139 generations, so while the 212 estimated by the genetic data alone is too high, it is not wildly off. This is when we are dealing with a fairly large group, and since the Clan Donald DNA project is the largest of its kind, other family studies will have fewer participants, and the estimates will tend to be more inaccurate. However, when you look at individual branches, the real problem becomes apparent. And it is precisely these smaller-scale issues that most genealogists I hear from look to genetics to solve.

Very frequently two men, who might share the same surname and whose Y chromosomes are similar but not identical, ask how long ago their common ancestor lived. The straightforward reply is to give an average figure, based on the same sort of calculations we have just used. But a close look at the Clan Donald genealogy shows how very misleading this can be for individual branches, even when, as we have seen, the results for the whole clan are reasonably close to expectations. You don't need sums to see this, just look at the diagram. Suppose the first two men in the bottom row asked just that question. We know from the genealogy that both men are descended from Ian Moidartach MacDonald (K), the eighth chief of Clanranald, who died in 1584, and that the two men, the current twenty-fourth chief (M) and a Mr. P. M. Macdonald (N), are separated from their common ancestor by eleven generations on one side and twelve on the other, making a total of twenty-three. Their Y chromosomes have accumulated a total of nine mutations, five on one line and four on the other since they last shared a common ancestor. This gives an estimate that they are separated from each other not by the twenty-three generations that we know to be the case from the genealogy, but by 9/(38 × 0.0031), which is 76 generations and more than three times as many.

At the other extreme, J. J. Macdonald (O) and Allan Douglas MacDonald of Vallay (Q) last shared a common ancestor in John, Lord of the Isles (E), who died in 1386. The genealogy shows the two men to be separated by thirty-nine generations, yet their Y chromosomes are only three mutations different. The estimate from the genetics puts the number of generations separating the two men at 3/(38 × 0.0031), which is 25 and far fewer than in reality.

Though I have chosen extreme examples I hope I have shown enough to convince you of the great inaccuracy of these estimates for individual cases. Not only do I believe they are next to useless, they are also misleading. Many people naturally think that increasing accuracy will come by increasing the number of markers tested. It will not. The random nature of mutation and the complications of parallel mutation will see to it that no estimate of the time to a common ancestor for any two men will be much more than a shot in the dark, even with the abundance of markers that are now available. And it is a fallacy to think that using more markers is the answer. Even with a perfectly respectable thirty-eight markers, any treatment of the genetic results on their own would appear to prove that J. J. MacDonald (O) and Allan Douglas MacDonald (Q) were far more closely related than Ranald Alexander MacDonald (M) and P. M. Macdonald (N), whereas the reality is the complete opposite. Caveat emptor.

8

The Jews

Genetics has been welcomed with open arms by European Americans. At the same time as clearing up various historical uncertainties, it has become almost a routine tool for genealogists. In little more than a decade the field has made the welcome transition from academic laboratory to widespread public participation, most of it self-funded. While the contrast between the enthusiasm for genetics of European Americans and the understandable suspicion shown by Native Americans is stark, there is one group of Americans who might be forgiven for being extremely cautious about the application of genetics to questions about their ancestry and identity. Given the racial and eugenic underpinning of their persecution by Nazi Germany, not to mention centuries of more general anti-Semitism, it would be unsurprising if Jews avoided any involvement with modern genetics at all costs. In fact the opposite is true. Although they are certainly aware of the possibility of its misuse, the Jewish community, especially in the United States, was among the first to embrace genetics, and for one very practical reason. Jews in both the United States and Europe have a high incidence of an

Star of David.

otherwise rare genetic condition known as Tay-Sachs disease. The effects of the disease are devastating. At first infants develop normally but after about six months a relentless and ultimately fatal set of physical symptoms makes its appearance. Gradually the child becomes blind and deaf and is unable to swallow. Muscles waste away, and paralysis gradually sets in. Death eventually occurs around the age of four. It is a heartbreaking condition for both the victims and their families.

The underlying biochemical problem in Tay-Sachs is the gradual accumulation of the waste products of fat metabolism, called gangliosides, in nerve and brain tissue. In normal babies these are cleared rapidly by the enzyme hexosaminidase A. In Tay-Sachs sufferers not enough of this enzyme is produced, and the gangliosides gradually build up until they begin to interfere with the proper function of the cells and the first symptoms of the disease appear. The ultimate cause of the enzyme deficiency is a genetic mutation in the hexosaminidase gene, a deletion of four DNA bases that not only removes a vital amino acid but also changes the amino acid sequence of the enzyme downstream of the deletion. This is the mutation that most commonly causes Tay-Sachs disease among Ashkenazi Jews, but there are now more than a hundred separate mutations known in different parts of the world, each interfering with the proper working of the enzyme. This raises the question of why the disease is so relatively frequent among Ashkenazi Jews and so rare elsewhere. One in twenty-seven Ashkenazi Jews in the United States is a carrier for Tay-Sachs, that is they have one normal and one mutated copy of the gene, compared with one in 250 of the non-Ashkenazi population.

The answer comes in two parts. Ashkenazi Jews have always lived as tight-knit communities and have tended to marry other Ashkenazim. In this respect they have something in common with other religious groups that have kept themselves physically apart and married within their communities—for example, groups like the Amish of Indiana and Pennsylvania. One consequence of this way of life is that married couples can both be descended from the same ancestor, even without knowing it. If this ancestor carried the Tay-Sachs gene, it can be carried down the two separate lines of descent to both husband and wife. Although everyone has two copies of the hexosaminidase gene, only one needs to be working properly to provide enough enzyme to clear the gangliosides. So, each with one normal and one mutant version of the gene, the parents are both carriers, but they do not have the disease. However, the basic rules of inheritance mean that there is a 25 percent chance, for each pregnancy, that the child will inherit two mutant versions. When that happens there is not enough enzyme, and the child is doomed.

This partly explains why Tay-Sachs is common in Ashkenazi Jews, but exactly the same argument would apply to other endogamous groups, like the Old Order Amish, among whom Tay-Sachs is unknown. Instead the endogamous traditions of the Amish have had genetic consequences in a much higher incidence of two other inherited disorders, Niemann-Pick and Gaucher's. Interestingly the residents of two adjacent non-Amish communities, also in Pennsylvania, whose ancestors emigrated to America from Germany and Switzerland through Philadelphia in the early eighteenth century and who have been largely endogamous ever since, have a high incidence of Tay-Sachs. But why is Tay-Sachs and not another disease with the same inheritance pattern, like cystic fibrosis for example, found among the Ashkenazi Jews? This is a much harder question to answer.

The evolutionary reasoning behind the high frequencies of recessive diseases is that although the sufferers who carry two mutant versions of the gene are at a great disadvantage, and in Tay-Sachs they never reach reproductive age, the carriers, with one mutant and one normal version, have some sort of compensating advantage. If they did not, then the mutant gene would soon disappear as every time a sufferer dies, two copies are lost from the population. Although this admittedly theoretical argument applies to all recessive diseases that are at all common, in only one case is there a watertight explanation. This is not a disease that affects Ashkenazi Jews but another community in America. These are the African Americans, and the disease is sickle-cell anemia. We will go on to see how African Americans have dealt with it, but for now let

The team: Richard, Bryan, and Ulla.

The *Mayflower* replica, Plymouth.

On Boston Common.

'Atticus Finch" in John Hancock's chair.

The Treat Rotunda at the New England Historical Genealogy Society's headquarters in Boston.

All aboard the California Zephyr.

Bear Lodge/Devil's Tower, Wyoming.

With Serle Chapman by the Tongue River, Bighorn Mountains.

The climb to the Medicine Wheel, Bighorn Mountains.

Custer's tombstone at the Little Bighorn battle site.

Buffalo in Yellowstone.

A steam vent in Yellowstone, gateway from another world.

Smoke from the Arnica fire drifts across Yellowstone Lake.

The Zephyr pulls into Emeryville,
California, at the end of its journey.

From sea to shining sea: the Pacific
at Ano Nueva, California.

Ulla at San Francisco Bay.

Bryan coming to grips with American
football.

Sea lions relaxing at Pier 29,
San Francisco.

The colonnades at Stanford University.

The Weatherford Hotel, Flagstaff.

Outside Stanford University.

On the way to Second Mesa.

Ulla in Utah.

The shadows lengthen over Monument Valley.

Hopi land meets modern times.

With Bennett Greenspan at the Houston headquarters of Family Tree DNA.

Union Station, Washington, D.C.

Ulla and the buffalo warrior, National Museum of the American Indian, Washington.

With Rev. Mark Thompson on the air at Sirius XM Radio.

With Dr. Henry Louis Gates Jr. outside the American Museum of Natural History, New York.

With Dr. Gates at the NEHGS Annual Dinner, Boston.

Job done: relaxing by Central Park, New York.

us explore why this particular genetic disease is so frequent within that community. The answer lies not in America but in Africa.

Sickle-cell disease is caused by a mutation, not in hexosaminidase but in one of the genes for the red-blood-cell protein, hemoglobin. In common with Tay-Sachs, only people with two copies of the mutant gene develop the disease. Since it too is fatal, what can be the compensating advantage enjoyed by carriers? The answer lies in their resistance to malaria, an often-fatal disease that is endemic in many parts of sub-Saharan Africa, especially in West Africa, to which many African Americans trace their roots. It has been shown that the red blood cells of sickle-cell carriers are more resistant to the parasite that causes malaria, even though exactly why is still not known. For some reason the parasites find these red blood cells harder to get into, which is normally the first stage of the infection.

What can the carrier advantage be in Tay-Sachs? Here there are only theories, but like sickle-cell, the reason has to be a thumping good one and probably involves resistance to an infectious disease. It doesn't have to be a disease that is still around. Even if malaria were to be eliminated tomorrow, the sickle-cell gene would still be there and store up problems for future generations. So the answer could lie in resistance to an infectious disease of the past. The likeliest candidate for Tay-Sachs is tuberculosis, which was endemic in the crowded urban ghettos that Ashkenazim were forced to live in for so much of their history. Although there is no direct evidence that carriers have a greater resistance to tuberculosis, and it is hard to imagine carrying out direct experiments to prove this, it is not a bad working hypothesis.

Whatever the reasons for the high incidence of Tay-Sachs among the Ashkenazi Jews, once the genetics were understood, there was a good chance that something practical could be done about reducing the number of babies born with this invariably fatal condition. Although a cure would be a great achievement, prevention would be better and also vastly more practical. The theory was simple: If carriers could be identified, they could either be discouraged from marrying each other or, more likely, never be introduced by the marriage brokers, though this is a profession that has now vanished among secular Jews. Also, carrier couples could be made aware of the one-in-four risk for each pregnancy and, in some cases, opt for prenatal testing and first-trimester termination of any affected pregnancies. No one likes terminations, but in many eyes they are so much better than bringing children into the world whose short lives will be ones of increasing suffering and pain.

It was the U.S. Jewish community itself that led the way, and the first carrier screening for Tay-Sachs began in 1971, based not on DNA but on a blood test for enzyme activity. Screening quickly spread to Jewish communities in Canada and Europe, and to Israel. Once the hexosaminidase gene had been identified, direct DNA testing for the mutant version began in the early 1990s using a simple mouth swab. By 2000 more than 1.5 million people had been screened, and more than fifty thousand (3.3 percent) of these were found to be carriers. The same DNA test that detected the mutant version of the gene in carriers was also applied to pregnancies where the parents were both carriers. That fact was usually established in couples who had already had a child with Tay-Sachs, and though they often wanted more children, they were understandably extremely eager to avoid having another Tay-Sachs child. The DNA was taken by a biopsy of the chorionic villi lining the uterus, which are part of the placenta and made up of fetal tissue. The biopsies were performed under local anesthetic at between eight and ten weeks into the pregnancy and with little risk to the fetus. By 2000 more than three thousand pregnancies had been tested, just over six hundred of which were found to carry two copies of the mutant gene. All but twenty parents opted to terminate the affected pregnancies, and of those mothers who declined, all went on to have Tay-Sachs babies. Of the more than two thousand pregnancies that continued after being cleared of having two copies of the mutant gene, only three went on to develop Tay-Sachs due to one or both of the parents having other mutant versions of the hexosaminidase gene that were not known at the time.

Led by the Jewish communities in the United States, the Tay-Sachs screening program was one of the first great successes of the practical application of genetic knowledge to an inherited disease. Among Jews in North America the disease has been effectively eliminated, although about ten cases a year are born to American non-Jews, who are not routinely screened. This is not unexpected for, as I hope you will come to realize, no race or ethnic group can ever be accurately defined by the genes they carry. That is not to say, of course, that it has never been tried. In the early days of research into Tay-Sachs disease its apparently exclusive occurrence among Jews certainly appealed to the eugenics movement of the early twentieth century as proof that Jews were a biologically separate race that was constitutionally prone to neurological disorders, as this surprising quotation demonstrates:

In the present state of knowledge of the etiology of idiocy and imbecility in general the only cause of their frequency among Jews that may be considered is the neurotic taint of the race. Children descending from a neurotic ancestry have nervous systems which are very unstable, and they are often incapable of tiding safely over the crises attending growth and development. They are often idiots or imbeciles.

The surprise is not so much the content, which is a fairly standard racist rant, but that it comes from an entry in the *Jewish Encyclopedia*.[1] The unexpectedly calm acceptance of the division of humanity into various different races was taken for granted in 1901 when this edition of the encyclopedia was published. It was written at a time when public opinion in the United States had turned against European immigration, especially from Russia and Eastern Europe. Leading opponents of Jewish immigration used the high incidence of Tay-Sachs disease as proof that Jews were an intrinsically inferior race incapable of adjusting to the American way of life: "The fact that Jewish immigrants continued to display their nervous tendencies in America where they were free from persecution was seen as proof of their biological inferiority and raised concerns about the degree to which they were being permitted free entry into the U.S."[2]

There are six and half million Jews living in the United States today, and although most of them trace their recent roots to Europe, the pattern of their arrival in America is distinct from that of other Europeans. Though they are a community bound together by a common religion, Jews have also thought of themselves, and been considered by others, to be a distinct racial or ethnic group, the distinction between the two terms being blurred in common usage. Within Judaism there are three main branches: The Ashkenazim, who make up the great majority of those living in the United States, moved there from central and eastern Europe, including Russia, during the nineteenth and early twentieth centuries, *Ashkenaz* being the medieval Hebrew for "Germany." The second group, the Sephardim, settled in Iberia for several centuries, and the name means "Hispanic" in Hebrew. Fewer in number than the Ashkenazim, the Sephardim in America settled earlier, many of them in the Southwest during the Spanish colonial period. The third main branch of Judaism, the Mizrahim, is a less geographically coherent group living mainly in North Africa and the Middle East. In comparison with the Ashkenazim and Sephardim, relatively few Mizrahim have settled in America.

All three branches of Judaism trace their origins to the Middle East two thousand years before the birth of Christ. Jewish tradition traces its ancestry back to Abraham, and as we shall see, there is some genetic evidence to support this genealogy. Abraham is also claimed by Islam as an ancestor of the prophet Muhammad, who lived in the seventh century AD and whose birth marks the start of the Islamic calendar.

As history informs us, the ancient kingdom of Judaea was overrun by the Babylonians at the start of the sixth century BC, at which time the First Temple, built by Solomon, was destroyed. The expulsions that followed the Babylonian victory marked the first of the many episodes of persecution and deportation that have punctuated the history of Judaism. The failure of two revolts against the Roman occupation, in AD 70 and AD 135, added to the Diaspora, as many of the population were scattered or sold into slavery. Yet despite this the Jews remained a cohesive group, and many of the exiled communities flourished. The Diaspora finally came to an end with the foundation of the State of Israel in 1948, though the term is now used to embrace all Jews living outside Israel.

The Ashkenazim emerged as a distinct group during the eleventh century, living in Alsace and along the Rhine, having arrived there from southern Italy after Charlemagne lifted restrictions on trade at the beginning of the ninth century. The period of stability for Jews that began with Charlemagne did not last, however. At the turn of the twelfth century the Ashkenazim suffered at the time of the First Crusade, when—in a classic example of the delusion of the masses—armed Christians set out for Palestine, the Holy Land as they called it, to reclaim it from Islamic rule. Less adventurous souls substituted the Jews as the enemies of Christendom, and began killing them instead of the harder-to-reach Saracens. Returning Crusaders, who had become deeply indebted to Jewish moneylenders, joined in the slaughter as an alternative to repayment.

To avoid the threat of massacre many Ashkenazim moved east to Poland, Lithuania and Russia. More expulsions followed from England in 1290, France in 1394, and again from Germany in the fifteenth century, all of which further increased the number of Jews living in eastern Europe. Jews were also the victims of mob violence, often condoned or even instigated by the authorities. Pogroms, as these acts of organized violence came to be called, occurred throughout the nineteenth and early twentieth centuries in countries like Poland, Russia, and Ukraine that had previously welcomed Jews. The numbers killed in pogroms ran into the thousands and triggered the beginning of large-scale immigration into the United States as Jews realized that they had a much safer and brighter future in America than in Europe. By 1880 more than a quarter of a million had already arrived in the United States, mainly from Germany, and these numbers climbed as a further two million from Russia and Eastern Europe arrived between 1880 and 1914. Terrible though the pogroms were, they pale in the face of the horrors of the Holocaust, which cost the lives of six million Jews during the Second World War. Tight restrictions on immigration imposed by the United States, Britain, and other nations limited the number able to escape persecution by the Nazis in the 1930s, but the great majority who survived the Holocaust eventually made their homes either in America or the newly founded nation of Israel.

On the face of it, given their troubled history of relocation and fragmentation over at least two millennia, the prospects for finding any common genetic ancestry among the Jews seem remote. Not only that, the search for any identifiable Jewish genes might seem an unwise "hostage to fortune" against future persecutions. And yet, as I have already mentioned, Jews have been among the most enthusiastic advocates of genetic research, not just for its medical aspects but for exploring their ancestry as well.

The first direct evidence that the Jews might have a separate genetic identity came from a study of the Cohanim, the hereditary priesthood of the Levites. According to tradition, the male descendants of Aaron, the brother of Moses, were chosen to serve as priests. The idea of testing this biblical tradition directly is attributed to Karl Skorecki, a Canadian physician and an Ashkenazi Jew himself, who thought of it when praying in a synagogue in a congregation with Sephardim. If the tradition was true and the patrilineal inheritance of the Cohanim from Aaron had been maintained over the three millennia since he lived, then they should share a Y chromosome inherited, ultimately, from Aaron himself.

Skorecki teamed up with a genetics research group from London, and between them they collected DNA from more than three hundred Jewish men from Israel, Canada, and the United Kingdom. Astonishingly, given the three millennia that have elapsed since Aaron's time, the London team identified a common Y chromosome (and its close mutational derivatives) in nearly 70 percent of Ashkenazi and just over 60 percent of Sephardic Cohanim, largely identified in the study by the surname Cohen. Among other Jews the frequency of this chromosome was around 10 percent, so this was a very significant finding indeed. The genetic age of the shared Y chromosome was an uncannily accurate three thousand years. It looked as though this really was Aaron's Y chromosome.

The Y chromosome shared by the Cohanim effortlessly assumed the mantle of a "Jewish" gene, although of course it will have been drawn from the pool of Y chromosomes circulating in the Middle East at the time. Possession of Aaron's Y chromosome does not prove descent from the brother of Moses, or that the bearer is Jewish. (This is like the familiar example of a fallacy: As all cats have four legs and all tables have four legs, therefore all cats are tables.) Many Christians and Muslims will also carry the same Y chromosome as Aaron, something worth noting in these troubled times. Indeed, since both Aaron and Muhammad were patrilineal descendants of Abraham, all direct male descendants of the prophet will be among them.

While the patrilineal descent of the Cohanim is spectacularly consistent with biblical tradition, it is something of a special case because of its association with the rules of inheritance governing the priesthood. It does not mean that there will be a general genetic similarity between Ashkenazim that might underlie their sense of common ethnic identity. However, equally amazingly, it is there to be found. We have already seen the indirect evidence from the much higher incidence of Tay-Sachs disease among the Ashkenazim along with the suggestion that it may have conferred an evolutionary advantage on carriers. But even if that were the case, it would only come about if the Ashkenazim did not often intermarry with their non-Jewish neighbors and grew in numbers from a small founding population. At some point in their history the Ashkenazim must have experienced a "population bottleneck," in the jargon of the professional geneticist.

The evidence from Y chromosomes and from mitochondrial DNA shows this to have been the case. One indication is the variety of different Y chromosomes. Whereas all the main clusters are found among the Ashkenazim, a reflection of the large and diverse population of the Middle East from where they were originally drawn, their frequencies are quite different from the surrounding populations of Germany, Poland,

and Russia. It is also important that the variety of different-detailed Y-chromosome signatures is also very much reduced in comparison with their non-Jewish compatriots. The same is true for mitochondrial DNA. All seven clans are there, but the frequencies are different. For example, roughly 8 percent of Europeans belong to the clan of Katrine, or haplogroup K, but among the Askenazim this rises to almost 40 percent. Within the Ashkenazi Katrines, as with their Y chromosomes, the number of distinct lineages is much lower than among Europeans in general. In fact, helped by complete mitochondrial sequencing, scientists have identified four founding mothers from whom 42 percent of all Ashkenazim are directly descended. Three are in the clan of Katrine and one in the clan of Naomi, or haplogroup N1, a very rare clan in Europe though commonly encountered in the Middle East. This is compelling evidence that European Jews have not homogenized with their European neighbors, and shows convincingly that the expansion of their numbers since the Middle Ages has been achieved by hereditary transmission rather than by conversion. The rule that a child born to a Jewish mother becomes a Jew appears to have been consistently followed. Compare European Christians, whose origins before the birth of Christ were exactly the same, where there is no inkling of any hereditary religious transmission, but plenty of vigorous, and often draconian, campaigns of conversion.

American Ashkenazim share the same genetic features as their European and Middle Eastern ancestors, and the same desire to find out more about them and make the link as European Americans do in general. I have been involved in many cases, and they all share the features of genetic explorations by non-Jewish Americans searching for their roots. One recurring theme, though, is the confirmation—or refutation—of a genetic relationship to Europeans whose surname is not the same. It was very common for German and Russian immigrants to anglicize their surname when they registered at Ellis Island or one of the other entry points to the United States. "Bachmann" became "Beckman," "Lehman" became "Lemmon," "Gottlieb" became "Goodlove," "Gutfreund" became "Goodfriend", and so on. This was never a completely consistent process, so seeking a genetic link with potential relatives, both in the U.S. and in Europe, by looking for a Y-chromosome match has become very popular among American Jews. But perhaps the most unexpected application of modern genetics, again using Y chromosomes, comes not from the American Ashkenazim but from their cousins the Sephardim.

Though Ashkenazi and Sephardic Jews share a common origin in Judaea, their paths through the Diaspora have been very different. While the Ashkenazim trace their European roots to central and eastern Europe, the Sephardim look to Spain and Portugal. Exactly how and when Jews began to arrive in the Iberian Peninsula is uncertain—it may have been as early as the incorporation of Hispania into the growing Roman Empire after the defeat of Carthage around 200 BC. The numbers grew with the Roman conquest of Judaea in AD 70 and the expulsions that followed the crushing of the Jewish revolts—the same dispersions experienced by the ancestors of the Ashkenazim. Until the adoption of Christianity by the Romans at the beginning of the fourth century, Jews in western Europe had largely avoided religious persecution, and even afterward the Christian church struggled to establish itself in the westernmost regions of the empire. The church's influence in Iberia collapsed completely early in the fifth century following the barbarian invasions and the rule of the Visigoths, during which Jews enjoyed two welcome centuries of freedom from religious intolerance. This changed with the conversion of the Visigoth king, Recared, to Catholicism in AD 587, which was followed by a century of persecution and expulsions. The Muslim invasion of AD 711 was welcomed by Jews and ushered in a long period of tolerance and prosperity that led to its epithet as the "golden age" in Iberia. This period of religious freedom came to an abrupt end as fundamentalist North African Berbers took control at the end of the eleventh century. Many Jews left Spain altogether rather than face the choice of death or conversion to Islam.

Gradually Spain and Portugal were brought back under Christian rule as cities and provinces were reconquered from the north. Although that rule was initially tolerant, the same religious fervor that energized the Crusades swept through the regained territories, and many Jews were killed. Things got steadily worse, and they faced the same dilemma as their cousins had to the south: conversion or death. Or if not death then humiliation and penury. Except that this time it was conversion to Christianity rather than Islam.

The final stage began after the surrender of the last Muslim emirate of Granada to the Catholic monarchs Ferdinand and Isabella in January 1492. Despite promises of protection, by March of that same momentous year, Ferdinand and Isabella issued the "Edict of Expulsion" that compelled all Jews to leave Spain by the end of July. Estimates vary, but at least 150,000 are thought to have left. Most settled around the Mediterranean, while the fortunate ones reached the welcoming embrace of the Muslim Ottoman Empire, mainly around Istanbul and northern Greece, prompting the sultan to send a sarcastic message thanking Ferdinand for enriching his country by impoverishing Spain. On October 12 of the same year Columbus landed in the New World.

Not all Jews left Spain and Portugal, where a similar conversion/expulsion order was issued in 1497. Those who remained chose conversion to Christianity, but often without a genuine religious realignment. While outwardly Catholic, many *conversos*, as they were known, continued to practice their true faith in secret. But they were not safe from the Spanish Inquisition, set up by the crown in 1478 to maintain strict adherence to Catholic orthodoxy and with the specific intention of ensuring that *conversos* remained converted. The Portuguese Inquisition followed in 1536, along even stricter lines. The courts set up by the Inquisition encouraged people to submit evidence against neighbors whom they suspected of continuing to practice Judaism. Even such indirect evidence as the lack of chimney smoke on Saturdays, indicating that the household was honoring the Jewish Sabbath, was used to root out heretics. The Inquisition routinely used torture to extract confessions. Those found guilty of heresy could confess, but those who refused were burned at the stake. By the time the reign of terror was reaching its peak, the newly established Spanish and Portuguese colonies in the New World were seen by *conversos* as safe havens, free from the dangers of the Inquisition, and many left Iberia for Brazil, New Castille (now Peru), and New Spain—comprising Mexico and Spanish possessions in the Caribbean and what is now the southwestern United States. This was the beginning of the Sephardic settlements in America, predating by three hundred years the mass immigration of the Ashkenazim in the nineteenth and twentieth centuries. Many *conversos* felt safe enough to begin openly practicing their hidden faith, but this was only an interlude, and by 1569 both New Spain and New Castille had their own inquisitions.

Genetically speaking, the Ashkenazic and Sephardic Jews are very different from each other, and this must be a reflection of their separate histories since the first Diaspora. The high frequency of Tay-Sachs disease, for example, is confined to the Ashkenazim. Though the Sephardim do suffer from several inherited disorders, including Tay-Sachs, these are no different in type or frequency to other Mediterranean peoples. For example thalassemia, the Mediterranean equivalent of sickle-cell disease that is caused by mutations in hemoglobin genes and offers a degree of protection to malarial infection, is found equally commonly in Sephardic Jews and non-Jews. Research using Y chromosomes also shows that, with the exception of the Cohanim chromosome, there is a much higher level of integration with adjacent European populations among the Sephardim than among the Ashkenazim. There is very little evidence in the Sephardim of the severe population bottlenecks that led to the current low level of Y-chromosome diversity among the Ashkenazim. The same is true of mitochondrial DNA, where there are no dramatic increases in the membership of any one clan compared to the Katrine effect among their Ashkenazim cousins.

Nonetheless the connection between the Sephardic Jews of Iberia and their origins in the Middle East is still there. A recent study comparing Y chromosomes of self-identified Sephardic Jews living around the Mediterranean (and thus descended from exiled Iberians) with the modern population of Spain and Portugal showed a very high proportion of integration, with almost 20 percent of the Iberian population judged to be of Sephardic/Middle Eastern ancestry.[3] However, as the authors explain, there are other possible reasons for this genetic correspondence, including the much earlier spread of agriculture from the Near East during the Neolithic period from about eight thousand years ago, and the maritime activities of Phoenicians from the eastern Mediterranean in the first millennium BC.

Even allowing for this ambiguity, the discovery in America of Y chromosomes with a likely Sephardic origin has had an enormous and sometimes life-changing impact on the men who carry them. I found this out by talking to Bennett Greenspan, the president of Family Tree DNA, whom we first encountered in chapter 6. When I visited him in Houston he told me that several of his Hispanic customers from Mexico and the southwestern United States were initially amazed that their Y chromosomes turned out to be "Jewish." Even though, as we have seen, this is an oversimplification, and no Y chromosome can be precisely defined as such, Greenspan assured me that in several cases this discovery awakened childhood memories of certain family customs that, until then, had no explanation—childhood memories like being told, "Never eat pork or you will die," of Grandma lighting candles on a Friday night, of throwing out an egg if it contained a speck of blood, or of covering mirrors when somebody died. "Now it makes sense," was a frequent response. These are the practices of observant Jews, and though the families were overtly Catholic, the Jewish traditions had lived on in secret. Genetics had revealed to these men that they were descendants of the *conversos*.

Their ancestors certainly had a lot to fear from the activities of the Inquisition. Even political authority did not spare them, as in the chilling case of Luis de Carabajal, which Greenspan related. In 1579 this Portuguese adventurer and slave trader was appointed governor of a new province in northern Mexico, Nuevo León, bringing with him his wife, Doña Isabel, and their children. All was well until one day Luis heard his wife talking to the children in a strange tongue and, thinking she might be going mad, consulted with his priest. The priest realized that they were speaking Hebrew and reported this to the Inquisition. Doña Isabel was tortured until she confessed and implicated the whole family in these forbidden practices. They were all imprisoned. Luis de Carabajal himself, despite being the one who inadvertently exposed the heresy, died in prison. The others were not so lucky and were burned at the stake in Mexico City. The residual fear of these and similar horrors still lingers among the descendants of the *conversos*, and it takes time for men who discover their genetic link to come to terms with the news. But when they do, Greenspan told me, several have taken the plunge and formally converted to Judaism.

Though the Sephardic and Ashkenazi Jews living in the United States today have very different histories and very different reasons for coming to America, their reaction to genetic research is in stark contrast to that of Native Americans, as the case involving the Havasupai Indians and the University of Arizona testifies. While Native Americans, as we shall see, generally regard genetics with extreme suspicion, American Jews have been at the forefront of genetic research. Bennett Greenspan is a good example. Himself an Ashkenazi Jew, the company that he founded was supported by the Jewish community in Texas right from the start. Indeed, as he explained to me, he deliberately sought their reaction even before he launched it in 2000, figuring that their traditional intelligent cynicism would highlight any potentially costly flaws in his business plan. While the very practical benefits accruing from Tay-Sachs research have virtually eliminated the disease among American Jews, there must be another element to account for the difference between the Native American and Jewish reactions to genetics. While both have reason to fear racial discrimination and the ever-present threat of eugenics, their reactions are poles apart. The main difference, as I see it, is that while Native Americans have been on the receiving end of gung-ho and invasive academic projects run by other people, Jewish Americans have looked into their genetics themselves.

9

The Africans

The last continental group to converge on America was the one with the planet's deepest ancestral roots, for it was in Africa about two hundred thousand years ago that our species *Homo sapiens* first emerged from among the several other contemporary hominids. They, too, had evolved from earlier forms, moving further and further away at each step from their more apelike antecedents. The hominid line, of which we are currently the only known survivor, converges with the common ancestor we share with our closest animal relative, the chimpanzee, about six million years ago. Some of our early ancestors, notably the very sturdy *Homo erectus*, left Africa at least two million years ago and spread across Europe and Asia, though not—as far as we know—to America. Some one hundred thousand years ago, a few *Homo sapiens* ventured beyond the continent of their ancestors to settle in Asia and in Europe. Slowly they replaced other hominids, like the Neanderthals, so that by thirty thousand years ago they had a virtual monopoly. Small pockets of related species lived on for a few thousand years in remote locations, like the Indonesian island of Flores, and the Altay Shan mountains of Siberia. Even now, regular though always inconclusive reports emerge from distant lands of strange semihuman creatures that are said to be the last survivors of other hominid species. Sasquatch and Bigfoot in the dense Pacific Northwest forests, the yeti of Nepal and the *migoi* of Bhutan in the Himalayas, and the almasty of the Caucusus Mountains. In the past I have investigated scraps of skin and hair from some of these without, yet, finding any convincing DNA proof of the legends. But that is another story. For the purpose of *DNA USA*, we can take it that humans evolved in Africa and spread from there to the rest of the world. So to say that we are all Africans has a literal truth even if it has become a rather tired aphorism.

Most African Americans, as is universally known, came to America as slaves. But they were not by any means the first Africans to leave the continent in historical times. Africans have, contrary to common belief, been coming to Europe for thousands of years. They were, for example, prominent members of the Roman occupation force in Britain after the invasion of AD 43. The troublesome Scots and Welsh "insurgents" soaked up Roman military resources, and by the second century it has been estimated that there were nearly three hundred thousand auxiliary troops in Britain, of which between 10 and 25 percent were Africans. This proportion grew during the British campaigns of the Roman emperor Septimius Severus in the early second century AD. He had a Roman mother and a Libyan father, though there is no compelling likeness of him as black. Black African soldiers also feature on Trajan's Column in Rome, the richly carved celebration of the eponymous emperor's victorious military campaigns erected in AD 155. Interestingly, the Africans are wearing their hair in dreadlocks.

Although intermarriage between the occupying forces and indigenous British women was not permitted by Roman law until AD 197, this only legalized what had been going on since the early days. At the end of their twenty-five years of military service, auxiliaries were given land and a pension, and many chose to remain close to their garrison with their families rather than return home. Some would have been African auxiliaries. The Vikings, who dominated large parts of Britain from the eighth to the fifteenth centuries AD, certainly brought Africans as captives to Dublin, Ireland, and Orkney, off the north coast of Scotland. Black Africans also feature in British mythology, especially in the legends of King Arthur and the Knights of the Round Table. Sir Palamedes, the original Dark Knight, was black. Although King Arthur was largely an invention of the twelfth-century Welsh cleric Geoffrey of Monmouth, his imaginative *History of the Kings of England* showed, if nothing else, that black Africans were no strangers to Britain and the rest of medieval Europe. These records contradict the later notion that, to Europeans, Africa was terra incognita, the unknown and unknowable land, the "Dark Continent," the last refuge of Kurtz in Joseph Conrad's *Heart of Darkness*. The discontinuity with past knowledge is summed up by the casual remark to me by the Dartmouth professor of English Gretchen Gerzina. Searching in a well-known London bookshop for a history of black people in Britain she was met with the sharp response from the saleswoman, "Madam, there were no black people in England before 1945."[1]

The large-scale commercial transatlantic traffic in slaves had its beginnings in the middle of the fifteenth century when the Portuguese began to trade extensively on the coast of West Africa. These early contacts culminated in the construction of a permanent trading post, Elmina Castle, in present-day Ghana, in 1497. Ghana's colonial-era name, the Gold Coast, prior to independence from Britain in 1957, describes the early ambitions of the Portuguese to find the precious metal in the African interior. Slavery was not the original purpose nor, contrary to some modern assumptions, did it start with the arrival of the Europeans. Since the twelfth century there had been a lucrative export trade in West African women wanted as domestic servants in North Africa and Arabia. What the Europeans did do was to create a demand for African slaves in the New World that, at least in the early days, complemented the trade with the Muslim world by favoring men. Neither was the picture painted by Alex Haley's influential book *Roots* of white traders rounding up Africans deep in the interior an entirely accurate one for it was often ruthless African rulers who caused their own people, or captives in war, to be led in chains to the trading posts on the Atlantic coast.

A great deal has been written about the inhumane conditions of the Middle Passage, when Africans were taken first to Brazil or the Caribbean and then, after "seasoning," as the process of acclimatization was euphemistically called, sold on to tobacco and cotton plantations in the United States. So much so that I need to provide here only the briefest summary of the economics. For more than a hundred years before they began to arrive in North America, African slaves had been taken across the Atlantic to the Portuguese and Spanish colonies in South America, particularly Brazil, where they were forced to work the sugar plantations. The early British colonies in North America, beginning with Jamestown, Virginia, in 1607, very nearly failed and were saved only by the cultivation of tobacco. The original plantation workers were not African slaves but indentured servants, mainly poor immigrants from England, though including a few Africans, along with some enslaved Native Americans. After their contracted term, by which time they had worked off the debts incurred by the cost of their passage, they were free to settle and even helped to do so with "freedom dues," which included land, seed, tools, and guns. Not only did these workers have to be replaced but they began to compete with their former employers and the price of tobacco started to slide. Matters deteriorated even further for the landowning elite in the 1640s when increasing numbers of immigrants survived their years of indenture. They responded by increasing the term, so that more would die before completion, and halted the grant of land as part of the dues. Their other solution was to look to Africa as a source of labor. By the late 1660s slavery had been legalized, and the terrible process of transformation to the centuries of despair had begun.

As the European powers, principally the British, French, Dutch, Spanish, and Portuguese vied with one another to control Atlantic trade to the New World during the mid-1600s, several African kingdoms expanded their territories, displacing hundreds of thousands of people and, in so doing, helping to create more slaves for export. This trade became highly regulated, with the relations between European slavers and the African merchants being on an equal footing. Licenses were granted and rent paid for port facilities. Slave raiding by maverick Europeans was rare and brought swift retribution in the form of confiscations and the withdrawal of licenses and other privileges.

Although the African slave trade has had an enormous influence on the demographic and genetic composition of America, the United States was by no means its principal destination. In recent years a great deal of careful work has been done to uncover the numerical details of the trade. The most comprehensive of these is the initiative by the W.E.B. Du Bois Institute of African and African American Studies at Harvard, which has collated the details of about 70 percent of voyages. Combined with other sources, we now know such details as gender ratio, survival rates, and ethnic origin. The figures are mind numbing. There were a total of around 54,000 voyages between Africa and the New World involving the shipment of more than 11 million slaves. By far the greatest number, 4 million, were taken to Brazil, mainly by the Portuguese. Mexico, Cuba, and other Spanish colonies imported 2.5 million, the British took 2 million to their Caribbean "possessions," and the French 1.5 million to theirs. All told, the plantations on mainland America took somewhere between four and five hundred thousand African slaves.

Although the slave trade began in the Portuguese colonies around the Gulf of Guinea, and West Africa remained the focus of the export trade throughout, it spread out in both directions to Senegal to the northwest and south to Cameroon, Congo, and Angola, and then all the way around the Cape of Good Hope to Mozambique and Madagascar.

We also know about the resistance: African slaves did not go quietly. There are records of some four hundred mutinies on the Middle Passage. Thousands of slaves committed suicide by jumping overboard. Nor was the trade completely without vocal critics. Two African kingdoms—Djola, in modern Chad, and Balanta in Guinea-Bissau—banned the trade within their borders and tried to suppress it in neighboring states. Some religious organizations, both Christian and Muslim, objected to the Atlantic slave trade on moral grounds. Indeed it was an Islamic scholar who, as early as 1614, formulated the very first legal argument to undermine its legitimacy. But these were mere whispers of disapproval compared to the deafening roar of commerce.

Numerically speaking the transatlantic slave trade peaked in the eighteenth century then diminished and finally ceased in the nineteenth. The Thirteenth Amendment to the Constitution technically abolished slavery in the United States in 1865, after the Civil War, but as we shall see, it continued in one form or another for many years beyond that. Brazil was the last country to abolish slavery, in 1888. By its end the slave trade had forcibly transported 11,313,000 Africans, the vast majority to the Americas. The first U.S. census, in 1800, put the total population at almost 5.5 million, of which 1 million, mostly American born, were classified as "black," a proportion of just over 18 percent. By 1860 the huge influx of overwhelmingly white Europeans had swelled the total population of the United States to 31 million, but the black population of almost 4.5 million still made up 14 percent of the total. Pretty clearly from these figures, African DNA was prospering in antebellum America despite the horrors of the slavery system.

Passenger lists from the ships bringing European immigrants to America are a vital resource for their descendants. Even though the work of the Du Bois Institute and others has cataloged the voyages of the slave ships, there is one crucial difference between the two. While the manifests certainly recorded the numbers, they were completely anonymous. African Americans almost never know the names of their ancestors aboard the ships that brought them to America. Their identity was unimportant to the traders in human cargo, and for generation after generation their descendants were left wondering if they could ever know who their ancestors were and where they had come from. This is a privilege that most European Americans take for granted. At least there are records available, even if they do not always yield all the answers. But for African Americans they are just not there, and this is where DNA has been able to help. As we have seen, European Americans have embraced genetics as a new tool with which to elaborate their own family histories, often finding they can pinpoint the name and the home of their own ancestors in Europe. I think this is wonderful, but it is a luxury that African Americans just do not share.

However, DNA can be, and has been, extremely valuable in overcoming the centuries of silence, and can reconnect African Americans to their ancestral origins. I first saw how and why when I was filming a short documentary for British television about a woman who was searching for her roots. An Afro-Caribbean, her name was Jendayi Serwah, and her parents had arrived in Britain from Jamaica in the 1950s. Since she was a small girl growing up in Bristol, in southwest England, Jendayi had always had a very strong attachment to her African roots. She had traveled to Ghana and gone through a ceremony to change her name to Jendayi, and she always wore African clothes. As part of the film I had been asked to analyze her mitochondrial DNA.

Unsurprisingly, to me anyway, she was in the clan of Lubaya, one of the thirteen African maternal clans to have been identified at the time. It was a very clear result. Searching my database, which was comparatively small back in 2001 when we were filming, I found a close match with a Kenyan Kikuyu whose results had been published in an academic paper in 1996. Documentary makers, reveling in the element of surprise that has become the staple diet of reality shows, love what they call "the reveal." In this case the plan was that I should not tell Jendayi what I had found until the cameras were rolling. I have since learned to be wary of this technique and refuse to go along with it if I suspect the genetics news may be unwelcome. But I was completely unprepared for what came next. I was not at all surprised that Jendayi's DNA had come from Africa. That is what I expected, given that it had traveled down the maternal line. So I gave her the results in a rather matter-of-fact manner, expecting a "Thank you, that's interesting" sort of response. Far from it. Jendayi was absolutely overjoyed. Tears of happiness ran down her cheeks as she gave me a big hug. I was dumbstruck—after all, what else had she expected?

The cameras were packed away and Jendayi left with the crew. The film went out a few weeks later, but still I couldn't forget Jendayi's response. I called her at home in Bristol and arranged to go and see her. What was it that had given rise to such an emotional response? It was this: She explained that her knowledge of her African roots had been secondhand, something that was taken from history books. Not that she didn't believe these histories, even though they were almost always written by whites, but they were always generalized and impersonal. What the DNA result had given her was proof of a quite different kind. Instead of having to rely on vague histories written by others, she carried a message from her ancestors in every cell of her body, in her own DNA. It was a secret document, a talisman of her African roots carried across the Atlantic in a slave ship by a black woman, her ancestor. Passed on through generation after generation of mothers and daughters to her own mother and then to her, it had slid unseen past slave owners who had strived to erase any sense of individual identity from her ancestors. And now her secret document could be read for the very first time. Its meaning was released, and she could see—almost taste—the savannah and the forest of her ancient homeland.

Africa, as we know, is the cradle of humanity and the ancestors of every man, woman, and child on the planet began their journey under an African sun. Humans have lived in Africa far longer than anywhere else, and so their genes have had longer to evolve and change. As a direct result of this antiquity, there is more variety among African DNA sequences than in the rest of the world put together. Since genetics is fundamentally about variety, African DNA is an especially rich confection, a real treat for geneticists like myself. Once again, it is mitochondrial DNA that comes out on top as the clearest window into the deep past, as it does in all indigenous populations, because it is the history told by women alone, untroubled by the frequently erratic behavior of men.

As we have seen already, the first detailed research on mitochondrial DNA of any indigenous people concentrated on Native Americans, mainly because the labs concerned were based in the United States. As we have also seen, within a region mitochondrial DNA falls naturally into a relatively small number of clusters, each with its own founder, the clan mother, and her matrilineal descendants. The early start of research into Native Americans meant that their mDNA clusters claimed the first four letters of the alphabet, A–D. By the time the first comparable studies emerged from African volunteers, the cluster notations had progressed through the alphabet to L.[2] This has meant that, despite African mitochondrial DNAs being by far the most varied anywhere in the world, they all belong to just one supercluster, L. Within L are three clear subdivisions that elsewhere would have had their own separate alphabetical notation and that thoroughly deserve their description as superclans. According to the internationally agreed classification, these three superclans are given the uninspiring monikers L1, L2, and L3. Such is the range of different mDNAs in Africa that each of these three superclans has its own substantial clusters. L1 is divided into five clusters, L1A–E; L2 into four, L2A–D; and finally L3 into seven clusters, L3A–G. If you can stand it, even these are broken down into L1A1, L1A2, and even L1A1A. Now we are entering the world of the phylogenetic aficionados, and we must tread warily because each step takes us in further away from real people and into the shrouded domain of the theoretician.

I have in the past attempted to humanize this excruciating yet necessary process by giving names to the clan mothers of each group, much as I did with the seven major European clusters and others around the world. After all, they were real individuals, and they lived long enough to have at least two daughters who lived long enough to have at least one daughter of their own. One of the qualifying conditions for clan mothers is that they must have at least two daughters. Being the most recent common ancestor of the whole matrilineal cluster, a woman with only one daughter would have a child who lived more recently than she did. Since my rule of thumb is to choose a name beginning with the cluster's

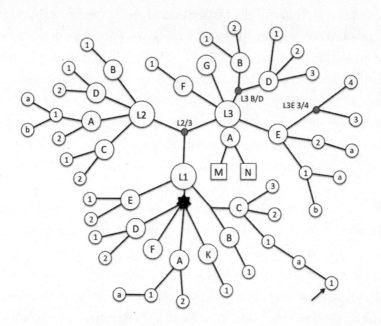

Figure 3. Maternal family tree of Africa. Circles are clusters of individual African mDNA sequences with connecting lines representing mutations. Cluster letters follow the international nomenclature—for example, the arrowed cluster is L1C1a1. The star marks the position of "Mitochondrial Eve," the root of the maternal tree. Gray circles are nodes with no extant sequences. M and N are the founder clusters of all mDNA found outside Africa. *Redrawn from A. Salas et al., American Journal of Human Genetics 71 (2002), 1082.*

alphabetical notation, all the African clan mothers begin with *L.* They are listed in the appendix. I would use them here rather than their alphanumeric equivalents, but having tried it, I found that the narrative gets even more confusing than it already is.

By 2002 the collection of samples from many different parts of Africa had reached the point where the maternal family tree could be drawn with little anxiety that it would need any major revisions, and I have drawn a simplified version in Figure 3. It certainly doesn't look simple, but it is the best I can do without losing all pretensions to scientific accuracy. On the tree, circles are particular mDNA sequences, and the interconnecting lines represent the mutations that have occurred between one sequence and the next. For those of you who are strong enough to follow these mutations through the family tree, I have put them in the appendix.

The first feature of the African maternal family tree that strikes home is its delicious complexity and that is because there has been more time for it to grow in Africa than anywhere else. The geographical location of the many different clusters and subclusters is also very revealing, not just about ancient events in the continent's history but about more recent known migrations, and about the effects of slavery. At the very root of the tree is a woman from whom we are all maternally descended. She lived about 170,000 years ago, most probably in East Africa. How do we know this? You are by now familiar with the principle that mutations accumulate in DNA over time, which is how we estimated the ages of the major Native American clusters. If we do the same kind of calculation for African mitochondrial DNA, the time it has taken for all the complexity seen in the tree to have accumulated is around 170,000 years, which is how we arrived at the date for our common maternal ancestor. There may be room for adjusting this figure if, for example, the mutation rate were slightly different, but it will not be far off. Fortunately, unlike with the equivalent estimates for Native American and European mDNA cluster ages, we are saved the complication of trying to identify the various founder sequences that have arrived from elsewhere. Since our ancestors started out in Africa, they did not come from anywhere else.

We also know that it all began with one woman whose existence was first realized in 1987 in a classic *Nature* paper by Allan Wilson and his colleagues from the University of California in Berkeley.[3] Using the much cruder methods of DNA analysis of the time, they showed that all reconstructions of the human maternal family tree led back, by an inevitable logic, to just one woman. Almost as inevitable was her instant nickname, "Mitochondrial Eve." Even though she has been dead for many millennia, we are able to reconstruct her mitochondrial DNA sequence by working backward from the branches of the maternal tree to its root. No one alive today has her sequence, since after such a long time, all her lines of descent have experienced several mutations. We know she is at the root of the family tree because her sequence, even though it is a reconstruction, is the closest of all human mitochondrial DNA sequences to our nearest living primate relative, the chimpanzee. It is also the sequence that is closest to mDNA recovered from the remains of our nearest extinct human relatives, the Neanderthals.

However, though we are certainly all descended from Mitochondrial Eve, she was not the only woman alive at the time, and the best theoretical estimate is that she shared the planet with five thousand others. But she is the only woman to have matrilineal descendants living in the present day. The others, or their maternal descendants, either had no children or had only sons whose mitochondrial DNA was not passed on. In succeeding generations these maternal lineages became extinct one by one, leaving Eve's descendants as the sole survivors. Starting with her nearest neighbors on the family tree in superclan L1, their present-day geographical settings reveal a good deal about how humans spread throughout Africa and out into the wider world.

The San Bushmen of southern Africa have a remarkable history. They number about ninety thousand and now find themselves in three modern countries: Botswana, Namibia, and Zambia. Among the last true hunter-gatherers, they live in spiritual union with their surroundings and with their prey, mainly the antelope of that parched region. Their language, Khoisan, is like no other, with a repertoire of sharp clicks that is almost impossible for outsiders to imitate. The San are on the edge, having been displaced from much of their former territory by diamond mining and tourism interests. In a reminder of the American government's treatment of Native Americans in the nineteenth century, the San have been forcibly removed from their traditional lands into newly created "settlements." They have naturally objected to this and, helped by supporters from around the world, have several court cases pending. Also, the United Nations Human Rights Council has criticized the government of Botswana for preventing their return.

The majority of the San are in the two clusters of L1D and L1K, which are among the most diverse in Africa. This alone gives an indication of the great antiquity of the San, but the presence of a small percentage of L1D among the Bantu speakers of southeastern Africa is a reminder that the San were once far more widespread and have been absorbed into the great flow of the Bantu agricultural expansion. This is one of the most intense and sustained intracontinental migrations not only in Africa but anywhere in the world. However, unlike in Europe, where the diffusion of farming from its origins in the Near East was largely a cultural transmission of agricultural skills, both linguists and geneticists agree that in Africa there was indeed an overwhelming migration of people.

In Africa as elsewhere in the world, agriculture allows a given piece of fertile land to sustain a higher population than does the hunter-gatherer lifestyle of the San and others. Farming began in earnest in Cameroon and Nigeria about five thousand years ago, and this is where the Bantu expansion began. It is a familiar story as pressure on space stimulated an expansion and, about 3,500 years ago, the farmers began to spread east and south into the lands of the Bushmen. Then as now the dense equatorial forest of the Congo was a formidable barrier, and the expansion slowly skirted its northern and western fringes. Agricultural settlements appeared around the African great lakes of Victoria, Turkana, and Malawi to the east of the forest, while the western route was largely coastal, probably by boat as the offshore island of Bioko (formerly Fernando Póo), in the Gulf of Guinea, was one of the first to be settled.

Bantu, the language of the farmers, is very different from that of the Khoisan-speaking hunter-gatherers, and it belongs to a completely different family. Where there was open conflict between the Bantu and the San, and with the Biaka and Mbuti Pygmies of the central forest, the iron weapons of the expanding agriculturalists proved superior to the stone of the indigenous hunters, just as European rifles were a crucial factor in the progress of the Indian wars in America. Those San who were not killed were either displaced to their current range, the Kalahari Desert in Southwest Africa, or absorbed into the Bantu settlements. They took their mitochondrial DNA with them, but in a tellingly familiar reflection of the way of the world, not often their Y chromosomes. As usual, it was the women who were "absorbed," not the men. While the San group L1D, with L1K, makes up nearly three-quarters of the Bushmen's mitochondrial DNA, only about 5 percent of Bantu speakers are in that clan, so the "absorption" was numerically relatively minor.

I am well aware that this kind of description, of phalanxes of Africans on the move with all the trappings of a grand plan, is far removed from the reality. There was no plan and no way of implementing one. All I have given here is a crude factual summary, lacking in any emotion. The expansion of the Bantu speakers happened at a time way before the days of organized armies or command from any central authority. Small bands of people moved slowly, taking with them the seeds and tools of agriculture. Plots of land were cleared and planted. Encounters with the indigenous hunters may sometimes have been violent, although the experience in Europe was relatively peaceful, and many of the hunter-gatherer native Europeans picked up the ways of cultivation from the incoming farmers but also carried on hunting.

At this stage they had no domesticated animals, but by the time the Bantu reached the great lakes of the East African Rift Valley, they had learned the ways of the pastoral people living in what is now Kenya. As the expansion turned toward southern Africa, it looks from the genetics as if they were joined by East Africans. You can see this by the way that another of the L1 clusters is geographically distributed. The cluster L1A is found in both East and Southeast Africa, and if we use the accumulated diversity as an indicator of its age, it probably started in East Africa about forty thousand years ago, well before the Bantu expansion. Two branches then evolved, one of which, L1A2, moved into the equatorial forest, where it is still found among the Mbuti of the Democratic Republic of Congo and the Biaka of the Central African Republic. The other branch, L1A1, is found throughout Kenya and Ethiopia, where it is also very diverse. However, only a limited number of lineages reached Tanzania, Malawi, and Mozambique, farther south. The most straightforward explanation is that they found their way there much more recently during the later phases of the Bantu expansion. There are some members of the clan in West Africa, but that has a separate explanation that we will come to in a moment.

The other major branches within the L1 superclan have a completely different geographical distribution within Africa. Cluster L1B is concentrated in West Africa, being virtually absent from the east and southeast. There is some L1B in equatorial Central Africa, where about 5 percent of inhabitants belong to the clan. They are far more varied than in West Africa, which suggests that the clan began in Central Africa and then expanded into West Africa about thirty thousand years ago, judging by its age there. The same goes for the sister cluster, L1C, which is both older and, at 23 percent, more frequent in Central Africa than farther west.

While Mitochondrial Eve is the matriarch of the L1 superclan, making it the world's oldest, the two other African superclans L2 and L3, are still very old, and much older than any equivalents in other parts of the world. Of the four clusters within L2, by far the most frequent is L2A, to which a quarter of all Africans belong. Branching off are two rarer granddaughter clans, L2A1a and L2A1b, both of which have all the attributes of a recent origin in West Africa and are probably associated with the very beginning of the Bantu expansions on the Cameroon plateau. These mitochondrial lineages were then carried along to East and Southeast Africa as part of that mass migration. These two clans are at the very tips of the family tree, whereas the age of the main L2A cluster is very much older, at about 55,000 years. The age is similar in both East and West Africa, making it difficult to choose one as the origin of L2A. The eastern and western branches seem to have separated about 14,000 years ago, around the end of the last ice age, when everyone was on the move. The date of this separation is judged on the accumulation of mutations from shared founder lineages, but where it occurred is hard to decide. I am inclined to think a Western origin is the more likely given that the sister clusters L2B–D are found predominantly in West and Central Africa and not in the east. But we are dealing with events of the very deep past, where some degree of uncertainty is both inevitable and forgivable.

The sister clusters L2C, L2B, and L2D are more or less confined to West and Central Africa, with traces carried along with the Bantu expansion. All four clusters in the L2 superclan are very old, with ages ranging from 120,000 years (L2D) through 55,000 (L2A), to around 30,000 for L2B and L2C. Taken together, the L2 superclan makes up a third of all Bantu mDNA lineages in Southeast Africa, bringing the combined total with a West African origin to 44 percent.

The superclans L1 and L2 are found only among indigenous Africans, or those like African Americans with comparatively recent roots in Africa. Only descendants of the third superclan, L3, are found among indigenous people outside Africa, whether that is Europe, Asia, Australasia, or America. But before we look at these lineages, how is L3 distributed within Africa itself? This superclan is found at its highest frequency in East Africa, but is certainly not confined to that region of the continent. Even so, it makes sense, as this is the closest region to the Red Sea, the Indian Ocean, and the Near East from where our species ventured out of Africa about sixty thousand years ago on its way to settle the rest of the world. Those adventurous souls were all members of cluster L3A.

Outside of East Africa, clusters L3B, L3D, and L3F are predominantly West African in both origin and geographical distribution. Like other West African lineages that we have already introduced, many were carried into southern Africa by the Bantu expansion. The small sister cluster, L3G, is largely restricted to the Hadza foragers who live along the shores of Lake Eyasi in Tanzania, who share the characteristic Khoisan click language of the San Bushmen. Though L3G does not seem to have moved south with the eastern stream of the Bantu expansion, as we shall see it is found in African Americans, though not frequently.

The oldest, most complicated, and most widespread of all the L3 clusters is L3E. You may be glad to read that it is the last one we need to look at in any detail. L3E appears to have originated in Central Africa, where it is still common, around 45,000 years ago. The cluster is well represented in the Bantu of Southeast Africa, and there are Kenyan Kikuyu who are also members. L3E is rare in West Africa but surprisingly common among African Americans. The reason for this was a mystery until it was discovered that a third of Brazilians with African roots are members of the clan. This suggested the former Portuguese colony of Angola as the origin, and sure enough, recent research has found that L3E is very frequent there.

It has taken researchers a long time to unravel the complexities of the genetics of Africa, a complexity forged by the vast size of the continent and its long history of human occupation. But now we have a good idea how the ebb and flow of humanity have sculpted the genetics of today's Africans, and as we shall now see, how this knowledge has helped so many Americans reestablish their connections to the continent of their ancestors—connections so often severed by the cruelty of slavery over the sea.

10

"I Am a Zulu"

"I am a Zulu," famously declared the best-known African American woman alive today, Oprah Winfrey. Being Oprah, her words were flashed around the world. Literally, of course, she is not a Zulu but an extremely gifted television host, author, businesswoman, philanthropist, and campaigner. But her choice of words was very telling: She used "I am," not "One of my ancestors was." Oprah's declaration of her African genetic roots followed a genetic test of her mitochondrial DNA, and though it was not the first time an African American had made this connection, it was certainly the most celebrated.

Tracing their African roots through mitochondrial DNA has been enormously helpful to many African Americans. While European genealogists concentrate on their patrilineal genetic roots through the Y chromosome, often flowing side by side with a surname, the obliteration of genealogically meaningful names for African slaves means this avenue is closed off. There are other reasons, too, why the Y chromosome is perhaps the worst DNA for African Americans to use to connect with their African roots. Fortunately the quest for matrilineal ancestors is

A village in Mozambique.

helped by the fact that more than 97 percent of African Americans have inherited an African mitochondrial lineage from their mothers, their grandmothers, and back through the generations to the slave ancestor who crossed the Atlantic. Of course African Americans usually have many more African ancestors than this, but the precision of mitochondrial DNA and the special place that maternal ancestors hold in all our memories make this small piece of DNA both the most reliable and the most precious.

As I saw at firsthand with Jendayi Serwah, finding a match between her and a Kenyan Kikuyu was a transforming experience for both of us. The match was an exact one in that the dozen mutations that defined her sequence were exactly the same as the anonymous Kikuyu. Anonymous because, like all academic research projects, the volunteers are protected by confidentiality clauses in the ethical approval that all such studies must obtain before they begin. These anonymous studies have been essential for revealing the indigenous patterns of African mitochondrial DNA, as we have seen, although they will never directly identify an individual as a relative. Nonetheless, close or exact matches between African Americans and indigenous research samples can give clues to an ancestral homeland in Africa. But it is not completely straightforward, for reasons that we have touched on in the previous section. For example, although Jendayi is an exact match with a Kenyan Kikuyu, that does not necessarily mean that the woman from whom they both inherited their mitochondrial DNA was herself a Kikuyu. They are both in the clan of Leisha (L2A). Leisha lived about fifty-five thousand years ago, well before the subsequent migrations within Africa that scattered her descendants far and wide across the continent. The precise sequence match means that their common ancestor lived far more recently than Leisha herself and probably within the last few thousand years. That is an average figure based on the mitochondrial DNA mutation rate, but the range of possibilities is very wide. We also know that the clan of Leisha is found not only in West Africa, but also in southeast Africa as a result of the Bantu expansions, which is what probably carried the lineage from West Africa to Kenya. We know from the history of slavery that it is far more likely that Jendayi's ancestor was taken from West Africa than from Kenya, but we cannot be absolutely sure. But it is so much better than nothing.

Comparing the mitochondrial DNA of African Americans with their indigenous African cousins is also very revealing about the geography of slavery. At first the absence of matches was perplexing. Take the clan of Lalamika (L1C). Many African Americans are members of this clan, but among indigenous Africans whose DNA has been tested it is comparatively rare. For example, among all the indigenous Africans sampled until 2006, there were only nine. Seven were from Mozambique in southeastern Africa, another from the Cameroon mainland, and the last from the island of São Tomé a hundred miles offshore. But, among a much smaller sample of African Americans, there were sixteen who belonged to this branch of the clan. Here I am using "African American" in its widest sense, in that this number included two Mexicans, four African Brazilians, and three "White" Brazilians.

At the opposite extreme, in the sister clan of Latasha (L1D) there are sixty-one Africans among the indigenous volunteers and not, so far, any African Americans at all. This mismatch is also explained by the geography of slavery. The indigenous African representatives of the Latasha clan are made up of forty San Bushmen from the Botswana Kalahari, twenty Bantu speakers from Malawi and Mozambique, and one solitary member of the Turkana tribe from northern Kenya. There are none at all among the almost five hundred African Americans included in this study. The reason is simply that slaves were not taken from Botswana's Kalahari Desert or from Malawi, both of which are far from the coast. They were, however, taken from Mozambique, so I think it is only a matter of time before we do see an African American in the clan of Latasha.

These two very different portraits are partially explained by the mechanics of the slave trade. While most ships left Africa bound for the Americas from the west coast, this was not always the original source of their human cargo. The Portuguese colony of Mozambique, facing the Indian Ocean, exported over a million slaves, and African Americans in the clan of Lalamika probably started their journey from there, via the offshore island of Zanzibar that had been an active Arab slave port for centuries. But before they crossed the Atlantic, slaves from Mozambique were often "stored" on the Portuguese island of São Tomé, awaiting shipment. Most likely the ancestors of the two clan members from São Tomé and Cameroon were also from Mozambique but somehow managed to avoid being taken across the Atlantic.

The relatively few absolutely exact genetic matches between African American and indigenous Africans in this example is partially a reflection of the extreme intrinsic diversity of African mitochondrial DNA that has had so very long to accumulate its huge range of mutations. Outside the family, almost everybody else is different. But it is also due to one other important but often-neglected reason. Put simply, if a region has not been sampled, there will be no matches within it. The research results we have been examining are distilled from about twenty different projects. By the nature of academic research, these projects are limited in scope. They are constrained by the amount of funding available, by the opportunities and permissions to study particular tribes or particular regions and by the time pressure to publish the conclusions. Fortunately the results from these separate projects can be combined because the mDNA sequences are all directly comparable. But there are always regions where no research has been done. Either it is too dangerous, or permissions are hard to get, or stretched funding agencies are not eager to pay for yet another project that looks awfully similar to the last one— there are a host of reasons. Angola, for example, was not widely sampled until recently due to the long-running civil war from 1975 to 2002. After the war was over and genetics results were published, mDNA matches between indigenous Angolans and African Americans began to appear.

The strength of feeling connecting African Americans to their African cousins now has its ambassador, its go-between, in mitochondrial DNA, and thousands of African Americans have drawn great strength from the genetic testimony it carries. Though this phenomenon shares some of the ingredients of the deep-rooted desire of many Europeans to discover which of the "Seven Daughters of Eve" is their ancestral clan mother, it has the added dimension of revealing, if only dimly, a past that has been completely and deliberately hidden. It bolsters the feeling of identity that many African Americans feel for their ancestral continent. While there were hints of this in the early days, as my experience with Jendayi had shown me, it was only when public access to testing became widely available that the full extent was revealed.

Public participation by African Americans began in earnest in the United States on February 21, 2003, when Dr. Rick Kittles and his business associate, Gina Paige, set up African Ancestry. Rick is a physician and geneticist, now at the University of Illinois, and as well as his work on the genetics of prostate cancer he has been instrumental in developing ways of estimating the proportion of African DNA in individual genomes, in ways that we will come to later. He told me when I visited him in Chicago that he remembers the start date of African Ancestry very precisely, as he and Paige had to rush to launch the company during Black History Week. Being an African American himself, like Paige, he realized the potential in linking individuals to Africa using their DNA, and appreciated the intrinsic qualities of mitochondrial DNA to do this. But as we have seen already, the potential is severely limited without large numbers of indigenous African DNA sequences for comparison. In 2002 there were not that many available, so he set about collecting his own. This is where the conflict between the academic and commercial worlds began, and it is one that I am familiar with myself.

The natural inclination of academics is to make results freely available through publication in scientific journals. Indeed it is an obligation for anyone employed by a university. Academic research was the only source of mitochondrial sequences until a decade ago, and without it there would have been nothing on which to base any commercial service. And without that there would have been no widespread public access, and no one outside a university research department would have been able to have his or her DNA analyzed on request. I know some of my university colleagues shrink from the idea of commercialization, but I believe they are mistaken. It seems to me that academic research should do the trail-blazing, but after a while, when the rules are understood and the field settles down, it doesn't seem appropriate that public money or research charities should pay for all the implementation. Researchers should get on with something new, or explore avenues that are unsuitable for private funding.

Kittles and I met in his office at the University of Chicago, and we both soon realized that we had shared similar experiences in opening the academic closet and letting people see inside—"taking the *American Journal of Human Genetics* to the street," as he put it succinctly. Although Kittle's main career has been centered on cancer genetics, he started out doing a graduate dissertation in biological anthropology. Being African American he naturally wanted to focus on Africa, but there were just no samples. There may be many shortcuts in research, but there is no getting away from the fact that to look into African genetics you need African DNA. However, there was an ongoing research collaboration in the department with Finnish scientists who were looking into the possible genetic basis for their country's distressingly high rates of alcoholism. And for that reason the lab freezers were full of Finnish DNA. You couldn't get much farther away from Africa, on the equator, than Finland on the Arctic Circle, but Kittles relished the thought of using the then newly discovered tools of mitochondrial and Y-chromosome DNA, albeit in much cruder ways than nowadays. In a nutshell he found that roughly half of Finnish Y chromosomes had come from Asia, while the mitochondrial component was determinedly European. As is often the case, the Y chromosomes were much more limited in their genetic diversity, a sure sign that a few men had more than their fair share of children. It is a picture I recognized only too well from Celtic Britain.

After finishing his thesis, Kittles was able to focus on Africa when he got his first position at Howard University in Washington, D.C. Howard was founded in 1867, just after the end of the Civil War, with the express intention of admitting African Americans, a guiding principle it maintains to this day. The medical school, which Kittles joined, has a reputation for training doctors from Africa, so, as well as a good complement of African Americans, there was a constant flow of students back and forth. That was the means by which Kittles built up his collection of indigenous African samples. He soon found a match for his own mitochondrial DNA, which is in the clan of Lingaire (L2C), among the Hausa of northern Nigeria. His Y chromosome on the other hand is European. The natural assumption for an African American is that one of his enslaved maternal ancestors was impregnated by a white man, possibly her owner—an extreme version of proprietorial rather than aristocratic diffusion, in other words.

However, I was fascinated to discover that there could be another explanation. Kittles told me that along the coasts of West Africa he had found that between 5 and 10 percent of fishermen also had a European Y chromosome. Coastal fishermen often have a different gene pool than do their inland neighbors. This was picked up a long time ago around the coast of Scotland when the blood-group composition of fishing communities was found to be quite different from that of the inland population nearby, and much more similar to that of fishermen and -women from other parts of coastal Scotland. The explanation of the European Y chromosomes among West African fishing communities is more about the flow of genes around the coasts of the Atlantic. The same flow has brought African DNA to the islands off the west coast of Scotland, as I discovered in my British research. It means, of course, that some African Americans may have gotten their European Y chromosomes from fishermen on the Atlantic coast of Africa, and not from white men in America. Some, but not all.

Kittles is one of the many African Americans who have visited the homeland of their ancestors in so far as it can be identified through DNA.

A few years ago he traveled to the Jos plateau in north-central Nigeria to visit the Hausa, the African tribe with the closest mitochondrial DNA matches to his own. It was both a frightening and a reassuring experience, he told me—frightening because the Jos plateau, in common with much of northern Nigeria, is embroiled in a war between Christians and Muslims. The roads were terrible, there were no lights, and no one dared to drive at night. There were potholes everywhere and larger craters with cars and even buses sticking out of them. The most profoundly disturbing images, however, were the unburied human bodies that littered the sides of the roads. Approaching a Hausa township one day, the car was pulled over and the driver gestured that he had better get out. Kittles was immediately surrounded by a hostile crowd, dressed in the long robes of Islam. However, what could have been a very dangerous moment was suddenly transformed. Among the men who surrounded the car was someone he recognized. Standing there was a man who was the spitting image of his uncle Clifford, his mother's brother. They looked almost exactly alike. It wasn't his uncle, but the mutual sense of recognition brought about by their shared ancestry immediately defused the tension. Rick suddenly felt the warm sense of kinship. It was, as he told me, "a very weird moment". Not really scientific at all.

Not being an African American, I don't expect I shall ever be able truly to understand the strong feelings of not knowing my roots and the void that it leaves: knowing you come from Africa, but not which part. Claiming in your mind the whole of Africa until you get a call from the messenger in your cells and can return to your home and begin to shape your true identity. That being said, many African Americans realize only after they visit Africa that they have more in common with America than Africa and that they can never truly go "home" again.

But things don't always work out as expected. Around a third of Kittles's African American customers turn out to carry European Y chromosomes. How do these men take the news? Not always well. About half of them are angry and frustrated and call up the company. They get mad and they want their money back. They know the general history of their race, but they look black and they certainly feel black. "How can I have a European Y chromosome?" they ask. "There must be some mistake." So the tests get repeated, and the answers come back the same. By then a lot of them have had time to think about it and have quieted down. Maybe they have asked their grandmother on their father's side, and she may have said there was talk of a white ancestor in the family a long time ago. Always ask a woman about these things, I have learned. Women always know more than the men.

When African Ancestry started out, it wasn't easy to know how to respond to the intensity of the reaction. Had Kittles anticipated it, he told me that he would have engaged a psychologist from the start. He has one now, who helps to manage the responses to unexpected or unwelcome news. With the experience of the years, Kittles has come to realize that a client's response to a DNA test is a mixture of two things: motivation and expectation. Many of his clients have well-established oral histories, and when the genetics runs counter to these expectations, it gets hard.

I feel a lot of empathy with Rick Kittles because at Oxford Ancestors we occasionally come across dissatisfied customers who just refuse to believe the implications of their DNA results. One of the services the company offers is to give our estimate as to whether a customers's patrilineal ancestor was a Viking or a Saxon or a Celt. The "Tribes of Britain" test, as it's called, is based on a large genetic survey of Britain and Ireland involving thousands of volunteers. Just recently I spoke with a man who was so sure his ancestor was Celtic that, when the "Tribes of Britain" tests showed his ancestor was probably a Viking, he called us. "I am certainly not a Viking," he said before demanding his money back. (I secretly suspect that many of our clients would prefer to be descended from Vikings, but he clearly wasn't one of them.) Usually the office staff can smooth things over, but he was so insistent that I called him myself.

"I want my money back," he said immediately. When I asked why, he replied, "Well, you said I was a Viking." "Yes," I replied. "But I'm not a Viking, I'm a Celt." Thinking that sometimes you just can't win, I asked, "How do you know?" "Because I am dark haired and short." At that point I gave up, realizing once again that DNA always struggles to reverse the deepest of psychological perceptions or identity associations.

11

All My Ancestors

We have traveled this far by listening to the clear music of just two solo instruments, free from the background rumble of the genome. The sharp and precise notes of mitochondria have traced the echo of our maternal ancestors back tens of thousands of years, following the journeys of women. The ferocious, warring blasts of the Y chromosome picked out the erratic history of men. The rest of the genome has been silent in narrating the story of our ancestors, leaving us free to concentrate on the separate melodies of men and women. Now is the time to turn up the volume on the rest of our genome, sit back, and listen to the sound of the whole orchestra.

The fraction of our genome carried by the two principal soloists is tiny. Mitochondrial DNA carries only thirty-seven genes on its compact circle of precisely 16,569 bases. Although very much larger, at 58 million bases, the Y chromosome has fewer active genes than mitochondria, only twenty-seven, owing to its decayed and enfeebled state. The rest of the genome, on which we depend for virtually all our genetic instructions, is far larger again, with just over 3 billion bases spread over twenty-three, for the most part, healthy and robust chromosomes, containing about twenty-five thousand genes. Already you must feel how easily this could become a cacophony of different sounds, completely drowning the sweet music of our soloists. And you would be right, because interpreting the ancestral signals coming from the main bulk of our genome is far less straightforward.

For a start, because we inherit the DNA in our genomes from both parents and it is shuffled at each generation, it is almost impossible to tell which ancestor is responsible for passing on which segment of DNA. We all have two copies of each gene, but without testing our parents directly, we cannot tell which copy we received from which parent. And that is just our parents. When it comes to more distant ancestors, who cannot be tested, then it becomes virtually impossible.

Then there is the issue of the generation paradox. Just like us, both our parents have two copies of every gene. But they each pass on only one copy to us and so we, you and I, only ever get half of one parent's DNA and half of the other. What happens to the rest of it? Some DNA may be passed on to our brothers or sisters, but the rest goes nowhere. The generation paradox arises because, for every generation back in time, the number of our ancestors doubles, but we still only inherit the same amount of DNA. To clarify this let us choose a particular gene, beta-globin, that controls one of the subunits of hemoglobin in our red blood cells. Thinking about our four grandparents, we will have inherited one copy of the globin gene from one of our paternal grandparents and the other copy from one of our maternal grandparents. But that leaves two grandparents whose globin genes we have definitely not inherited. Going back another generation, to our eight great-grandparents, we have inherited our globin genes from only two of them, leaving six grandparents whose globin genes did not get through to us. Likewise, going even further back and still with only two globin ancestors at each generation, fourteen out of our sixteen great-great-grandparents, and thirty out of thirty-two great-great-grandparents, will not have given us our globin genes. We will never know, without a lot of extra work, which of these thirty-two ancestors once carried the globin gene that we have inherited.

The globin gene is only one of thousands, so even if we received our globin genes from only two of our thirty-two great-great-grandparents, we will have inherited the copies of plenty of other genes from all of them. However, because the number of ancestors keeps on doubling at every generation that we go back, there will come a time when there are ancestors from whom we don't inherit any DNA at all. But when will that be? With 25,000 genes and two copies of each, that makes 50,000 separate DNA segments. So when the number of ancestors exceeds 50,000 there must be some from whom we get no DNA. It is a simple calculation, just doubling at each generation 2, 4, 8, 16, 32, and so on. After fourteen generations this mathematical series gets to 16,384, and exceeds the 50,000 mark by generation 16, when we have 65,538 ancestors. With a generation time of twenty-five years, that is only 400 years ago, which takes us back to the beginning of the seventeenth century, about the time of the first English settlements in America.

However, this calculation assumes that we inherit our DNA in a neat and equitable way from our ancestors. In fact, which particular segments we inherit from which ancestors is completely random and therefore governed by the rules of chance. We get more DNA from some and less from others. This spread means that there are some much more recent ancestors, probably within only six generations, from whom we haven't inherited any DNA at all. With the same 25-year generation time, that is only 150 years ago.

While this is not so long back, the numbers of ancestors are doubling at each generation and growing at an alarming rate. Which is where the paradox shows itself, because at some point the number of ancestors will exceed the entire population of the world. What is the solution? It is this: Although the number of ancestors doubles at each generation, some of them will be the same people. Not our parents, obviously, but two of our grandparents could, theoretically, be the same person. It's unlikely but possible. The chances increase as we go back until, at some point, it becomes inevitable. Where that point is depends on how our ancestors lived. If they were mainly endogamous—that is, marrying among themselves like Ashkenazi Jews, for example—these "double" ancestors may have lived quite recently. For more exogamous ancestries, they will have lived further back in the past. However, whether from an endo- or exogamous ancestry, there will inevitably come a point when one person is the ancestor of everyone alive. This sounds absurd, but theoreticians have calculated that in an exogamous population of one million people, this person lived only twenty generations, or five hundred years, ago. Even when other factors, including a more realistic figure for the world population and the effects of migration and geographical isolation, were brought into the calculation in more sophisticated models, this "universal ancestor" still lived just seventy-six generations, or, assuming the same twenty-five-year generation time, only 1,900 years, ago.[1] I do find it astonishing that, compared with the quarter-million-year history of our species, one individual, from whom everyone alive today can trace a line of descent, lived so recently.

He or she was only our most recent "universal ancestor," and as you go further back in time the same theorists predict that the population divides into individuals from whom everyone can trace at least one line of ancestry—and the rest, who were the ancestors of nobody alive today. That point is reached at five thousand years ago, about the time the Pyramids were built. So the slaves who built them would either have been the ancestors of everybody alive today, including you and me, or of nobody. Beyond that point everyone is descended from exactly the same set of ancestors, though along different lines. Even further back, the proportion of people with no living descendants increases until only one couple remains who were the ancestors of everyone living today through every line of descent, except two. As you might expect, these estimates are surrounded by caveats. However, even the most sophisticated models incorporating factors like the opening of sea routes do modify the timing, but not by much. The principle is still valid, and the conclusion, strange though it may seem, is inevitable.

The two exceptions to this theoretical scenario are the direct matrilineal and patrilineal ancestries that are by now so familiar. They coalesce into universal ancestors a lot further back, around 65,000 years for the "Y-chromosome Adam" and 170,000 for "Mitochondrial Eve." Why the difference? It all has to do with the behavior of men who have more than their fair share of children, something I explore in depth in *Adam's Curse.*

Even without the complications of the fairly recent universal ancestors, our genomic DNA ancestry is a snarled tangle compared with the linear simplicity traced by mitochondrial and Y-chromosome DNA. We know exactly which ancestral path they have taken from the deep past to the present. They are the two clear voices above the tumult of the genome, but despite their clarity they cannot tell us the complete story of our ancestry, and a lot of it remains hidden. I used to think this was a blessing in disguise because it is almost impossible to grasp the concept and complexity of our complete DNA ancestry in a way that means anything. All narrative is lost as the numbers of our ancestors grow into the thousands and beyond. For a long time I could see no way of narrating our genetic past that went beyond mitochondria and Y chromosomes. I was happy listening to the crystal-clear voices as the Callas and Domingo of genetics sang their duets. I wasn't interested in the cacophony of the orchestra and chorus.

Then, by chance, I did catch something in the air: the faintest possibility of a melody rising above the swirl of countless different instruments. It came in late 2008, when a former colleague, John Loughlin, was explaining to me how a new DNA technology was making his work on the genetics of osteoarthritis so much easier. He had been looking for genes involved in the cascade of biochemical events that lead to severe arthritis, requiring a major joint replacement. This was a very reasonable quest because he had already shown that this particular form of osteoarthritis had a high hereditable component, *ergo* there must be genes involved. But as so many prospectors of the genome have discovered, it is hard work, and the rewards rarely match up to the promise first imagined—more like panning laboriously for specks of gold than striking a mother lode. I knew that he had spent many years teasing apart the genomes of hundreds of patients who had undergone joint-replacement surgery. He and his team were looking for differences between these patients and the hundreds of other people of similar age and background who did not suffer from osteoarthritis. Like other scientists on a similar quest, he used the enormous range of genetic markers that had been discovered en route to decoding the entire human gene sequence. The theoretical basis was that if his joint-replacement patients inherited particular versions of any of these markers, compared with the control group, then this might indicate the chromosomal location of an "osteoarthritis gene." It didn't mean that the genetic marker itself *was* that gene, but that it might lie nearby. The work involved was enormous, with each marker being either analyzed alone or with a few more. With thousands of them to get through, the vast majority of which would be duds, it was a massive effort, and John spent most of his time either raising the money to pay for the work or cheering on his research team.

The technical breakthrough that made the difference was arrived at independently by scientists in Britain and America: They developed ways of fixing DNA to glass. Since I had known one of the English pioneers, Ed Southern, who was working in Oxford, I had seen the early versions using sheets of window glass about ten inches square, which he covered with a matrix of small drops of DNA solution, each containing a different synthetic segment of DNA that had been made to match exactly its equivalent in the human genome. These glass sheets were early prototypes, and by the time John Loughlin started to use the new technology for his osteoarthritis research, the whole system had been miniaturized so that half a million markers now fitted onto a silicon "DNA chip" about one inch square. The matrix of synthetic DNA markers was now far too small to see the individual spots, so the reactions were observed under a microscope.

For John and the other gene prospectors, this advance meant that instead of examining the thousands of markers in his patients and controls, either individually or in small groups, he could analyze half a million at once with a single DNA chip. No wonder he was pleased. The chips were and still are expensive, and the machinery to read them is beyond the budget of most university laboratories, so it made sense for this work to be contracted out to commercial labs that could benefit from the economies of scale. So now all John needed to do was get hold of the DNA, send it off to one of these labs, get the results back, and interpret them— and spend even more time raising the money to keep going.

While I could not fail to be impressed with the technical achievement of the DNA chips and the sheer slog it was saving John and his team, I did not immediately see how this would help unravel the tangled ancestry of our genome. It was only when John referred me to 23andMe, a Californian company that was offering chip-based DNA tests to the public, that I began to understand their potential. On its Web site were examples of what the company called "chromosome paintings." The moment I saw them I caught the first melody from our genomic ancestors. What had been until then a formless noise, audible only to the oscilloscope of computation, suddenly resolved into woodwinds, strings, and brass. Within a week I was on my way to San Francisco.

Company headquarters are in the broad, winding avenues of Mountain View, right at the heart of Silicon Valley twenty miles south of San Francisco. As I found my way, I passed neat yet unremarkable two-story buildings set back from the road and half hidden by trees. The buildings may have been unremarkable, but the signs outside were certainly not, for here and in nearby Cupertino were the research headquarters of some of the best known global companies in electronics and computing: Google, Apple Computers, Cisco Systems, Siemens, and more. I had

managed to arrange a visit at such short notice because the company's
director of research was former Stanford geneticist Joanna Mountain,
whom I had met on the academic conference circuit and whose work
on mitochondrial DNA I knew well. The place was buzzing, because
23andMe had recently won the 2008 *Time* magazine Invention of the
Year award.

The central theme of the business was to use the DNA-chip technology
to provide customers with information about their risks of developing
a range of genetic diseases. Some diseases, like sickle-cell anemia and
Tay-Sachs, have a simple one-to-one correspondence between identifiable mutations in known genes and developing the disease. However, for
most diseases with an inherited component, like the osteoarthritis that
John Loughlin was researching, the links to specific genes are a great
deal more tenuous. One of the great research efforts of the past decade
has been to identify these genes, hoping that what was true for Tay-Sachs
and sickle-cell would also be true for diabetes, hypertension, Parkinson's,
and the rest. The initial optimism that drove the furiously competitive
search for these genes, fueled by the prospects of patenting them and
making a fortune, was soon tempered by reality. They proved to be at
first elusive and then quite impossible to tie down. As the British geneticist (and wit) Steve Jones once remarked, trying to find them resembles
T. S. Eliot's description of the hunt for Macavity the Cat, the "Napoleon
of Crime," in his *Old Possum's Book of Practical Cats.*

He's the bafflement of Scotland Yard, the Flying Squad's despair:
For when they reach the scene of crime—Macavity's not there.[2]

That is not to say that genes are not involved at all in these common
diseases, just that the prediction that they would be small in number and
big in effect turned out to be wrong. The reality is that the genes involved
are many in number and individually weak in their effect. Although the
search for the "Napoleons of Crime" may have been abandoned, there

are plenty of minor accomplices that have been found "loitering with intent" and taken in for questioning, and the outcome of these enquires has been to use the DNA results to adjust an individual's risk of developing a disease.

People's perceptions of risk are notoriously wayward, particularly when there are numbers attached, and bear only scant relation to anxiety levels. For example, I am much more worried about being crushed by a herd of stampeding cattle than I am of being killed in a car accident, even though the statistics show that I have it completely the wrong way around. In the United States 105 people were killed by cattle between 2003 and 2007 while 192,256 died on the roads. Not strictly comparable, I know, but you see what I mean. I could modify my personal risks downward by never going into a field full of cows, or avoiding traveling in a car. One of the presumptions of personal genetic risk analysis is that we will modify our lifestyles accordingly: If we have a higher than normal genetic risk of obesity then we will go on a diet, or if we are told we have an elevated risk of developing diabetes then we will avoid sugar. I have always thought this a very tenuous piece of reasoning. After all, millions of people smoke though they know very well that it might kill them. But I could not have put it better than a journalist, from *The New Yorker*, I think, who once interviewed me about the results of some tests he had done on himself through another company. After he had finished his questions, I asked him what he was going to do about this new knowledge about himself. "Eat more broccoli," came his sardonic reply.

As you can imagine, there has been a great deal of debate about the value of these results, and even whether the tests should be offered to the public at all. A lot of this has been among professional medical geneticists who are fearful that people will discover they are at high risk of developing a grave genetic disease. There are good reasons for taking this seriously, and during my time in medical genetics, I have been impressed with the arrangements for counseling people who are contemplating a DNA test because of a family history of a genetic disorder. None illustrates the dilemma better than Huntington's disease. This insidious and invariably fatal affliction seems deliberately designed to maximize cruelty to its unfortunate sufferers and their relatives. The symptoms of neurological and personality collapse do not show until around the age of thirty, after which there is a steady decline toward dementia and death. The pattern of its inheritance means that children who see one of their parents succumb have a 50 percent chance of inheriting the mutant gene and developing the disease themselves. Unlike Tay-Sachs and other recessive disorders, one mutant copy of the gene is enough to give the symptoms.

Finding the Huntington's gene, in 1993, was one of the triumphs of the early years of genetic exploration and immediately offered the prospect of a genetic diagnosis before the onset of symptoms. Not that anything could be done about stopping the development of the disease, but there were circumstances when the DNA test was requested, most commonly when someone who was at risk but too young to show the symptoms was contemplating starting a family. Often this was someone who had already witnessed the suffering of a parent but did not know whether they carried the same death sentence in their DNA. There are so many factors that need to be considered, even before having a DNA test, that professional advice is essential. How will you respond to a positive result? Or even a negative one, which you would have thought would bring unrestrained relief but is often met with a deep feeling of guilt. How about identical twins? Since their DNA is exactly the same, the result of a test would apply equally to both, but what to do when one wants the test but the other does not? It is no surprise that suicide has been the response of some to finding out that they have the mutant gene, and under these circumstances it is easy to see that it would be catastrophic to offer the Huntington's DNA test directly to members of the public without the backup of professional genetic counseling. I think considerations of this kind have made the medical genetics profession extremely wary about direct-to-consumer genetic testing for less acute disease susceptibilities, which is why on the whole, it is not in favor.

I have certainly had vigorous arguments with my colleagues about this, and I think they are wrong. First of all I think they underestimate the sophistication and common sense of customers. Second, their response is both arrogant and hypocritical in the sense that the same medical genetic community that has trumpeted the benefits of genetic research now wants to restrict public access. By all means root out the charlatans, but instead of sniffily looking the other way, help companies that have the resources and the motivation to do a difficult job well. And, by the way, I am not being paid to say this.

Although their primary objective is in the health-care aspects of modern genetic analysis, 23andMe was also well aware that the same genetic information could be used for personal ancestry research. Organizations like Oxford Ancestors and Family Tree DNA had proved that there was a market, while the appetite for personalized genetic health-risk evaluation by members of the public had never really been tested when they launched in 2007. No one knew how much people would be prepared to pay and how many would want it. But for ancestry the figures were there, and since it required no additional genetic analysis, only interpretation and presentation, it was sensible to offer ancestry testing as a sideline. And it was this sideline that brought me to Mountain View.

I was met at the door by Joanna Mountain and one of the cofounders, Linda Avey, whose background is in marketing. I had prepared a short presentation, mainly about the narrative qualities of DNA, as I recall, after which Joanna did the same, explaining how the company was adapting its DNA-chip system for ancestry applications. The others in the audience were mainly young, mostly scientists or software engineers. After a short tour and some individual meetings I left to rejoin the hell that is Highway 101 going north to San Francisco. Except this time I hardly noticed. I was very pleased with how things had gone. Although this was only an intial contact after all, I came away with a very positive impression of the company and the people and, more important, an offer of help with my research for *DNA USA*.

What had intrigued me from the start and had seemed to offer a way into the complexity of our genomic ancestry was the way in which the ancestral origins of human chromosomes were portrayed. Each of our twenty-two pairs of autosomes—that is to say all of our chromosomes, except the X and Y—were laid out in horizontal rank and in numerical order from the largest, number 1, at the top to the smallest, number 22, at the bottom. Each chromosome was sliced lengthwise along the middle so that the top and bottom slices represented the two copies of each chromosome that we possess. In examples of people with a mixed ancestry, different colors picked out the segments of their DNA that had come from one of three continental origins. Dark blue for European, green for African, and orange for Asian—which in the United States is a proxy for Native American.

This was not the first system to estimate the continental components in an individual's genome. An earlier method had been developed that gave a quantitative estimate of African, European, and Asian DNA, but it did not break this down into chromosome segments. Rick Kittles, the cofounder of African Ancestry, had used a system, called AIMs for "ancestry informative markers," with some interesting results that we will look at later, but something about the numerical brutality of AIMs made me wary of its use in individuals. Chromosome painting, on the other hand, seemed to overcome my misgivings and come much closer to the real situation for individuals with ancestors from different continents, and the visual representation made it much harder to misinterpret.

So how do you go about painting someone's chromosomes? The underlying science depends, as always in genetics, on the variations between one individual's DNA and the next. Without these there would be no genetics. One of the triumphs of the Human Genome Project, aside from reading the entire human DNA sequence, has been to discover millions of tiny differences between human genomes, known by their acronym SNPs, which we have already encountered in the Y chromosome. The initials stand for "single nucleotide polymorphism," which means a difference only in the DNA sequence at a particular location on a particular chromosome. For example, where the sequence might be GGATTA on one chromosome and GGATCA on another, this is a SNP. Millions of

SNPs have been discovered throughout the human genome, which once found can be identified by the DNA sequences on either side. So the SNP we introduced as GGATTA/GGATCA is flanked by unchanging DNA sequences that are known from the human genome sequence. You need only a sequence of around twenty bases on each side to uniquely identify any SNP. These short DNA segments are easy to synthesize, and easy to immobilize on a chip. Once on the chip they are able to detect which of the two sequences is present at the SNP in any DNA they are asked to test—or "interrogate" in the lingo. After some clever chemistry the spot on the chip where the synthetic SNP sequences are attached glows a fluorescent red for one version and green for the other, with these tiny signals being picked up by a powerful automated microscope. As there are half a million SNPs on a typical chip, and the microscope can scan all of them within a few minutes, very soon you know the sequence at all half million SNPs in the DNA being interrogated.

However, these chips are analyzing DNA from individuals who have two copies of each chromosome. This means that there are not two but three possible results for each SNP. If, in our example, GGATTA glows red and GGACTA glows green, when both chromosomes have the GGATTA version the spot will glow red. On the other hand, when both chromosomes have the alternative sequence GGACTA at the SNP, the spot will be green. But there will be times when both versions are present and one of the chromosomes has the sequence GGATTA while the other has GGACTA at the SNP. Under these circumstances the spot on the chip glows both red *and* green. Fluorescence filters on the microscope can deal with this and record both versions of the SNP. Even though the chip has analyzed half a million bases, this is only a fraction of the total of three thousand million. However, these bases have been chosen as the ones that are known to vary between chromosomes. We also know the precise chromosomal position of all half million of them.

At each generation our chromosomes shuffle their DNA sequences. Most of the time the chromosomes we received from our mothers and the ones we inherited from our fathers don't have much to do with each other. They lead physically separate lives in the cell nucleus and carry on barking their instructions to our cells quite independently from one another. In most of our body cells they live apart throughout our lives, but in our germ-line cells there is a final embrace. Just before our chromosomes become packaged into eggs or sperm, the pairs line up with each other and swap DNA. Then they move apart and go their separate ways into the germ cells. There is a very sound evolutionary reason for this tender parting, as it creates an enormous amount of genetic diversity in the next generation that protects the offspring from parasites and pathogens, again something I explored in *Adam's Curse*.

Whereas this was once thought to be a completely random process, and that DNA exchanges could happen anywhere along the length of the chromosomes, it turns out that this is not so. It now seems that there are "hot spots" along each chromosome where these exchanges are much more frequent. Rather than being completely random, as in a properly shuffled deck of cards, it is as though there are runs of cards that stay together.

The blocks of DNA between hot spots that are not disrupted by shuffling can be tens of thousands of bases long and contain several SNPs. This means that they tend to retain the combination of variants at each of the SNP sites within them. So a block with five SNPs might have the combination, as seen by the chip, of red/green/green/red/red, each one indicating the presence of a particular variant at the SNP site. This introduces a new level of discrimination, as there are now 2^5, or 32, possible combinations for this segment. This is the same principle that provides for the enormous range of genetic signatures generated by only a few markers on the Y chromosome when they are used in combination. Although the situation on the autosomes is far less helpful than on the Y chromosome, not least because exchanges are not exclusively confined to hot spots, the presence of relatively undisturbed segments of DNA is nonetheless valuable for the next stage in the chromosome painting process.

After the Human Genome Project finished in 2003, there were a lot of geneticists looking for something to do, and a lot of idle machinery. Some of them plowed on with sequencing other genomes, first mouse, then chicken, and so on. They are still going and, predictably, the species being sequenced are becoming more exotic. In 2011 the complete DNA sequence of the nine-banded armadillo and the canary were on their way to completion, in company with multitudes of potentially useful bacteria and fungi.[3]

Other geneticists switched their researches to studying the DNA variation among individual human genomes and soon began to realize that the human genome was falling into blocks. Thanks to the discovery of DNA-exchange hot spots and the cooler regions in between, a huge international scientific effort to describe these blocks as fully as possible began to take shape in 2002. How many there were, where the boundaries were, and so on. The impetus and the large sums made available were driven by the optimism of finding the elusive common disease genes, the "Napoleons of Crime." By knowing where these blocks were, it was going to be easier to locate these genes by the simple strategy of association between the blocks and the presence or absence of the disease in question in large numbers of patients and controls. Where the association with a particular block was high, then Macavity must be hiding nearby. Surely?

To discover how these blocks were behaving the HapMap Project (after "haploblocks," as these chunks are known) looked in detail at the genomes of individuals from three different parts of the world.[4] The chosen ones, 270 in all, came from Africa, Asia, and Europe, and each individual's DNA was typed for about 3 million SNPs. The ninety-strong African contingent was from Ibadan in Nigeria, members of the Yoruba tribe; the ninety Asian volunteers were from Tokyo and Beijing; while the ninety Europeans were actually Americans with their roots in northern and western Europe. The work was divided up among labs in the United

States and Canada, England, China, and Japan with each lab concentrating on different chromosomes, as they had in the initial sequencing of
the human genome. Many of the HapMap scientists were veterans of the
Human Genome Project and knew their way around their favorite chromosomes. From these results half a million SNPs that were favorably
placed within each block were selected, and these were put on a DNA
chip. Like the Human Genome Project, one attractive feature of Hap-
Map was the release of data into the public domain, and it was through
this release that software engineers were able to get their chromosome
brushes out and start painting.

The aim of chromosome painting is to assign each block in an individual's genome to one of the three continental origins represented by the
HapMap volunteers. This is of course a gross simplification, but it seems
to work. Let us take President Obama as an example, not that I have his
details. (I was told that the president has declined to be tested while he
is still in office.) I may not have his chromosome painting, but I have a
pretty good idea what it would look like. As everyone knows, President
Obama has an African father and a European American mother. He has
inherited one chromosome of each of the twenty-two autosome pairs
from his father, with DNA blocks that will likely match up with the Nige-
rian volunteers more than they do with the Asians or the Europeans who
helped build up the three continental reference collections. Equally, his
other chromosome in each pair has come from his mother, whose DNA
blocks will probably all match the European more closely than either
the Asian or African chromosomes. The painting software makes these
comparisons for blocks of DNA of about ten thousand bases all along the
twenty-two pairs of chromosomes. With a total of 3 billion DNA bases
to cover, this makes a total of about thirty thousand blocks to color in.

Mike MacPherson, the scientist who helped develop the program,
explained to me on my visit that there are six possible combinations
for each of these blocks along the chromosome pairs: African/African,
Asian/Asian, European/European, and then the combinations of Afri-

can/European, African/Asian, and Asian/European. Mike's algorithm chooses which of these combinations fits best with the DNA being analyzed and fills in the painting accordingly. "African" blocks are colored a light green, "Asian" blocks are orange, and "European" blocks are dark blue. As each block is analyzed and painted separately, there is a set of conventions that govern the coloring of the top and bottom slices. When both copies of a block have their best match with only one of the reference samples, as in African/African, for example, then both top and bottom slices are painted green. The rules come into play when the two blocks match different reference samples. So for African/European blocks the top slice is painted green for African, and the bottom slice is dark blue for European. An African/Asian block has the Asian orange on the top slice and African green underneath. The third mixed block, Asian/European, has blue on top and orange below. There are examples of all of these in the illustrations for chapter 19.

Returning to my theoretical reconstruction of the president's chromosomes, given that his father, Barack senior, was from Kenya and his mother, Ann Dunham, was a European American from Kansas, I would expect the chromosome pairs in his body cells to be African green on top, following the convention mentioned above, and European blue beneath pretty well as in the monochrome version in Figure 4 (A). Chromosomes don't actually look like this in real life. This is just a diagram of one of them, but it does give me the opportunity of pointing out one of the features of chromosomes. They are divided into two arms, separated in the diagram by the gray disc. The discs are there to represent attachment points for muscle-like proteins that help to pull the two chromosomes apart during cell division. Although the attachment points are made of DNA, their sequences are very repetitive and hard to analyze and consequently have not been included in the HapMap coverage or the SNP chips and are shaded gray in color versions. There are one or two other gray regions that have been left off the chips for the same reason, but they are only small and we can forget about them.

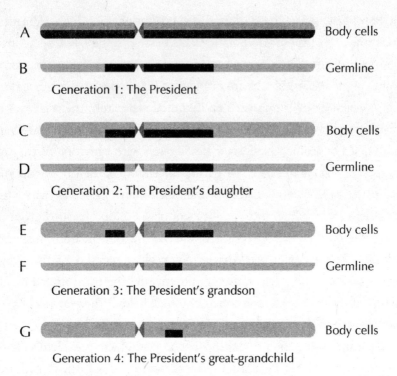

A ————————————————————————— Body cells

B ————————————————————————— Germline
Generation 1: The President

C ————————————————————————— Body cells

D ————————————————————————— Germline
Generation 2: The President's daughter

E ————————————————————————— Body cells

F ————————————————————————— Germline
Generation 3: The President's grandson

G ————————————————————————— Body cells
Generation 4: The President's great-grandchild

Figure 4. Following one of President Obama's chromosomes through four generations. The light gray blocks are DNA of African origin, while the darker blocks have a European origin.

The president's children have inherited one of each pair of chromosomes from him and the other from their mother, Michelle, the first lady. The chromosome coming from the president (B) is an amalgam of the two chromosomes in his body cells (A), shuffled by DNA exchange. There is usually only one exchange on each chromosome arm at each generation, so the chromosome going to his first daughter, Malia Ann, in Figure 4 (C) might look like this, although the random nature of DNA exchange makes the precise pattern unpredictable.

We know from conventional genealogical research carried out by Megan Smolenyak and reported in the *New York Times* on October 7,

2009, that Michelle Obama has some European ancestors. However, for the sake of simplicity, we will ignore these and assume that all her ancestry is African. So the example chromosome in Malia's body cells would look like C, with the mixed African/European chromosome (B) from the president and an African chromosome from the first lady.

Looking into the future, to the time Malia Obama has her own children, the chromosome she passes on will be another amalgam of the two she inherited from her parents, randomly shuffled by DNA exchange, like D in Figure 4 perhaps. If, to keep it simple, she marries a man with an African genome, her child—let's say it is a boy this time—will have arrangement E in his body cells. Most of the DNA in this pair of chromosomes has an African origin, all except for the European DNA in the dark blocks that have come, originally, from his great-grandmother, Ann Dunham. Sure, we can give a percentage of African and European DNA in this pair of chromosomes (roughly 88 percent African and 12 percent European by the look of it), and all the other chromosomes once we have "interrogated" them on the DNA chip. But in my view that doesn't take us much further than the ethnic ancestry tests derived from the AIMs. What really distinguishes chromosome painting from its forerunner is that, since we know precisely where genes are located on each chromosome, we can tell the continental origin of each one in any individual.

In the president's theoretical grandson—let's call him Harry—most of the genes along this chromosome will have an African origin, but for genes located within the two-tone blocks, he will be working on a fifty-fifty combination of African and European genes. If, for example, the gene for the ABO blood group was in one of these blocks, then his blood group will be decided by a mixture of African and European DNA. If the block contained a muscle protein gene, his muscles would be powered equally by African and European genes. Since both the size and boundaries of these blocks is so random, unless they happen to be identical twins, it is extremely unlikely that any two of the president's grandchildren will inherit the same blocks of European DNA, and hence the same European genes, on this chromosome. When all the chromosomes are brought into the comparison, then what was vanishingly unlikely becomes virtually impossible, and—though each of the president's grandchildren may have close to the average of one-eighth European DNA that is expected—the number and the identity of the genes with a European ancestry will be quite different in all of them.

By the time of the next generation, assuming once more for simplicity that Harry marries an African, his child, the by-now-former president's great-grandchild will have only one small segment of European DNA on the chromosome we have been following. It came originally from the president's mother and has survived through four generations, diminishing by roughly half at every one. It may survive for many more generations to come, or it may be eliminated by the forces of random chance at any one of them.

We have followed only the ancestry of one chromosome through three generations, and even then we have assumed that the chromosomes that joined the genealogy from outside are of entirely African ancestry. As you can imagine, where these incoming chromosomes are themselves built up of blocks of DNA with different continental ancestries, the picture soon becomes very complicated. But, however intricate it is, we would still be able to recognize the ancestral origin of the blocks of DNA and identify the genes that were contained within each one of them. I liked the way the chromosome portraits got so close to the actual situation and illustrated it so well. Our genomes are all mixtures built up of bits and pieces from a huge number of ancestors, and when these ancestors came from different continents, the variety is both obvious and intriguing. This was what I wanted to explore in America, and as soon as I returned to England from San Francisco, I began to plan in detail how to go about it.

SECOND
MOVEMENT

12

The New Englanders

I realized at the start that this was going to be very different from all the genetics projects I had ever done. Previously, in my research into the genetic history of Polynesia, Europe in general, and most recently Britain and Ireland in particular, I had started with the presumption that I needed to study large numbers of people. Partly that was in order to satisfy the statisticians that guard the entry gates to respectable scientific journal publication and for whom nothing is true unless it is also statistically significant. They miss an awful lot. Second, I didn't know from the start what was out there, and in limiting numbers I might have missed a vital genetic ingredient. This last point has turned out to be important, as I have been able to pick out the small proportion of people whose DNA seems completely out of place but which are the echoes of historical events. Events like the introduction of African slaves during the Roman occupation of Britain, or the Korean genes that have washed up on the shores of the North Atlantic. Very unusual DNA that might well have been missed in less substantial surveys. In America, however, this would have been impossible. It would have cost millions to use the same

Col. William Prescott, Bunker Hill Memorial, Boston, Massachusetts.

approach, and anyway would only replicate what others had already done. I also have to admit to being rather reluctant to write another genetic history along the same lines I had already followed in other parts of the world. Besides, there were issues I wanted to look into that would not have interested purely scientific publications.

The whole project began, in my own mind, to assume the character of the "last big job," much as bank robbers are said to look forward to just one final payoff before hanging up their weapons. And always getting caught. For my "last big job" I decided on a completely different approach. Instead of the comprehensive and detailed planning that had gone into earlier research projects, I decided this one would be guided by chance events from the start: I would just see where they led. I think I might have been influenced by seeing *Easy Rider* again for the first time in years in which the characters played by Peter Fonda, Jack Nicholson, and Dennis Hopper wander aimlessly around the southwestern United States on their way to New Orleans for Mardi Gras. Or maybe it was the image of "Shoeless Joe" Jackson materializing in the cornfields of Iowa in Phil Alden Robinson's movie *Field of Dreams* starring Kevin Costner. "If you build it, they will come." Freed from having to please the stats police, since I had no need or ambition to publish elsewhere than in *DNA USA*, I would take it easy, travel around, and just see what happened.

Though I had been to America on many occasions I didn't really have a sense of how big it really was. Like many Europeans, my experience had been more or less confined to the east and west coasts, with a couple of days in Chicago. Of course I knew it was big, but not *how* big. I knew it lay somewhere between the extremes of a country whose dimensions I was used to, namely Britain, and infinity. I didn't have any real feeling for where America was on this scale. To put this right, I decided to travel coast to coast by train, at least in one direction.

My son, Richard, just eighteen, agreed to come with me for the first leg east to west. Of all my research assistants, both paid and unpaid, Richard has been on more DNA expeditions than any other, and from a very young age. He was only six years old when he came with me to Scotland and toured the blood donor clinics. He was twelve when we set off for a three-month tour of Australia, New Zealand, and Polynesia. This time I knew would be the last. He was about to leave home for college. The three weeks we would spend in America would be the last of many long adventures together, after which I knew things would never be quite the same again. On the return leg across America I would be joined by Ulla, whose natural effervescence was to prove invaluable when it came to recruiting volunteers.

I would like to say that we set off that September with the easy confidence and optimism that are essential for any road movie remake. But that was far from reality. When Richard and I arrived in Boston it was cold, gray, and pouring with rain. City rain—falling not with anger or finesse, but just dripping from the sky as though a damp gray sponge had been suspended from unseen pinnacles low across the city. The sullen drops fell languorously onto the metal window ledge of the downtown hotel, tapping out a monotonous rhythm in 4/4 time. The next morning it was still raining. The task I had set myself seemed overwhelming and without limit, mocking my conceit that I could ever tackle such a gargantuan undertaking as writing a genetic biography of the United States of America.

By the next day the skies had cleared, and Richard and I set off for our first appointment. We walked across Boston Common, past the lake, where weary boatmen propelled their swan boats brimful of visitors around a figure-eight course. The first colors of fall were touching the elm trees, their farthest leaves a pale yellow under a bright blue sky. The sun, by now well above the horizon, reflected off the golden dome of the Massachusetts State House near the top of Bunker Hill. We were heading for the affluent streets off Commonwealth Drive and the headquarters of the New England Historical Genealogy Society in Newbury Street, the oldest and best-known genealogy society in America. We passed shops that signaled the discreet wealth of the neighborhood: Chanel, Valentino, Diane von Furstenberg, past carefully choreographed climbing plants, their tendrils twined around iron railings. The headquarters building of the society had once been a bank and still retained the formal grandeur of its earlier life. Great bronze doors lay behind a cluster of clipped green bushes, their austerity mitigated by a Visitors Welcome sign. Inside we found ourselves in a large oak-paneled hall, a sumptuous chandelier dripping from the high ceiling, with books and portraits of early New Englanders lining the walls. If there was ever a place to immerse myself in this region and its people, 101 Newbury Street was it.

As soon as Richard and I were inside, my doubts from the first days in Boston began to ebb. Here we could get down to some serious work. And so it proved. We could not have been made more welcome, and within minutes of our arrival, I was in the boardroom explaining my plans for the book to the society's senior staff. We were given a room in which to interview our volunteers and take DNA samples. It was perfect, and I began to make arrangements for the rest of the week. Although I had tried to keep prior arrangements to a minimum, for the sake of the road movie effect, there had to be some forward planning. I had asked the society whether any members had ancestors who had arrived in America before 1700. Of all the European Americans, this was the group I most wanted to meet and, if possible, to sample. Their ancestors had been among the first to arrive and had shared New England with the Indians, principally the Wampanoags, peacefully at first but not for long. If there were to be any genetic evidence that relations between them went beyond the aloofness of the historical record, then these were the people whose chromosome portraits might hold the answer.

Another topic I was eager to cover was the impact of genetic testing on genealogists since it became widely available ten years ago. I had assumed it had been useful, if only by the numbers of people who had used DNA to explore their past, but I had only rarely had the opportunity to discuss it with people who were genealogists first and foremost. I wanted to know more about the how and the where, and for this the society was the best place to start.

Two weeks before Richard and I arrived in Boston, the society had kindly sent out an appeal for volunteers among the membership through its weekly e-bulletin. The criteria were, I thought, rather demanding. I had asked for volunteers whose ancestors had arrived in New England before 1700 and who would be able to come to Boston during the week I planned to be there. I thought I would be lucky to get half a dozen replies, instead of which I got more than four hundred! Time alone meant that the short-listing had to be brutal, and I was forced to decline some fabulous offers.

Waiting for me were twenty chromosome-painting kits. Since the lab system required me to specify the name of the DNA donor in advance, and I didn't know who the volunteers were going to be, I had to invent a list of pseudonyms. Without a moment's hesitation I realized there was only one possible theme: Hollywood. After all, that is how most of the rest of the world has come to know America. I decided on characters, rather than the actors themselves, just in case casual readers might think I had dreamed up a devilish way to recover DNA from, or even clone, the long-dead stars of the silver screen. When I opened the box the kits were already labeled for "Rhett Butler," "Norma Desmond," "Holly Golightly," and so forth.* At first my volunteers could choose, and "Margo Channing," played by Bette Davis in the 1950 movie *All About Eve*, was the first to be snapped up, but by the end the choice was governed by which characters remained. Although the Hollywood pseudonyms were a product of the lab system and my own lack of close planning, I have gotten to know the DNA results through their movie names because that is how they were sent back to me from the lab. It was quietly amusing to receive them as attachments to e-mails beginning "Dear Sugar" or "Dear Atticus." That is when I realized that the response system was either completely automated or entirely staffed by the very young. I am keeping the Hollywood names for the narrative of *DNA USA*, although I have offered volunteers the chance to reveal their true identity should they wish to do so. And many did.

I had known "Margo Channing" for several years, having met her on one of the society's many organized visits to the UK. It was through "Margo" that I had first been introduced to the society when I gave a lecture at one of its annual meetings ten years ago. That was when the use

---

*Quotation marks used to denote pseudonyms throughout.

of DNA in genealogy was in its early infancy, and I think it may well have been the first time a genealogy society anywhere in the world had ever had a talk from a geneticist. And it was "Margo" who had made sure that the society welcomed me on this present visit. So she was the first of my volunteers or "victims"—as they began to refer to themselves as the week progressed. Like many of the society's members and all of my volunteers, she had been researching her New England ancestors for many years. This had taken her back to England, where she had found several families related to ancestors she had tracked through the records as having arrived in New England in the early seventeenth century. I had worked with her on some of these and confirmed a DNA connection for a couple of them. It was a pleasure to meet with "Margo" again a decade after our first encounter, looking as young as ever, and with the same razor-sharp mind behind penetrating bright blue eyes. "Margo" settled into one of the society's most comfortable chairs.

She began immediately with a detailed recitation of her New England ancestry. This she did entirely without notes, just like the Celtic *seannachies* I had heard about in the Scottish Highlands, whose job it was to recite the lineage of any new clan chief on his appointment. Though of course "Margo" had composed this history from detailed written records, she knew it by heart. To me it sounded very complicated, full of unfamiliar names, but to "Margo" it was second nature. I was at risk of being swamped with information, and I was very glad I had brought my voice recorder so I could disentangle it later at leisure. This became a frequent experience with all my volunteers from the society, and throughout the trip. I just could not pay full attention and take good notes at the same time.

"Margo" told me that her paternal grandmother had several *Mayflower* lines, so she was multiply qualified for membership of one of America's most exclusive clubs, the Mayflower Society, founded in 1897 and restricted to descendants of the Pilgrims who sailed from England and arrived at Plymouth, Massachusetts, in 1620. Her maternal grandmother, who was a Bigelow—another famous New England name— before she married was herself descended from Christopher Todd, who arrived in New Haven in 1637 with the Davenport company. An ancestor of "Margo"'s maternal grandfather was Thomas Savage, whose daughter Mary had married Capt. John Crocker, master of the *Cambridge*, who, "Margo" was once told, had a wife in every port. She is now on the track of this disreputable relative and is checking through port books in London and Boston to track his movements. Another of "Margo"'s ancestors was Anne Marbury Hutchinson, who made herself very unpopular with the Massachusetts Bay Company by teaching Scripture in her own home, so much so that she was told to go and join the renegade colony of Rhode Island. I felt myself embraced by New England history, recounted in living detail by someone whose ancestors had been here for a very long time indeed. The details were not so important as the experience of listening to the stories of lives lived long ago, told with such intimacy that "Margo" could have been talking about her relatives down the street.

One of "Margo"'s ancestors in particular caught my attention. Nicholas Meriwether had arrived in the Virginia Colony in the 1650s, but there was no information on his own father who was, presumably, from England. I had read about the Lewis and Clark expedition across America in the early 1800s and remembered that Captain Lewis's first name was Meriwether, with the same unusual spelling as "Margo"'s ancestor, so I asked her if she knew whether he was a relative. She had been unable to make the connection to Captain Lewis, but she did know all about him, naturally. He had been born in Albemarle County, Virginia in 1774, and the Meriwether name came from his mother, Lucy. His father's mother was also a Meriwether from the same valley. He died unmarried and, by "Margo"'s account, unappreciated. Although he became governor of the Louisiana Territory, he came in for a lot of criticism from the U.S. Congress at the time partly because he had built up large debts and partly because he had been very slow in filing his expedition reports to Congress. He did not die a hero's death but met his end, either by suicide or murder, in a Mississippi boardinghouse on the notoriously lawless Natchez Trace in 1809. (There had been plans to exhume his body to solve what had become a classic historical mystery, but early permissions to do so had recently been rescinded.) One of the fascinating anecdotes about Meriwether Lewis that "Margo" had at her fingertips was that among the medicines he had taken with him on the historic expedition across America was a colonic purge containing mercury. This bizarre fact was being used to locate the precise sites of his camps through traces of the metal in the soil.

"Margo" then told me that she had become involved in a DNA study of the Meriwether name. More than that, she had brought a summary of the results with her, which she proceeded to pull from the folder in her bag. I am glad to say that I am still close enough to the details of Y chromosome lab results to be able to read those charts and pick out unusual features. There were "Meriwethers," "Merewethers," "Merry-weathers," and a few other spelling variations. When I looked at the Merryweathers, there were several different Y chromosome signatures, indicative of separate founders of the name. This is not surprising given that the spelling makes its meaning clear and would likely have been chosen by several men in England when the name became hereditary around 1300. I then looked through the table of DNA results for the Meriwethers who had volunteered. Unlike the Merryweathers, all thir-teen Meriwethers had exactly the same Y-chromosome profile at every single marker. This had to mean that, unlike their more populist name-sakes, the Meriwethers were descended from one man in the United States. Unfortunately there were no English Meriwethers in the panel of volunteers for the simple reason that none could be found. The only Meriwether left alive in England was an elderly woman with no surviv-ing male relatives: The name has almost daughtered out. "Margo" expe-rienced a flurry of excitement a few years ago when she discovered the birth certificate of a Nicholas Meriwether who had been born in London only a few years previously. The thrill quickly evaporated when she dis-covered that he was the son of Dr. Will Meriwether, past president of the Meriwether Society, who had been working in England at the time. The search continues. The consistency of the genetic results means, I think, that all the American Meriwethers, whether tested or not, are descended from Nicholas. Or if not from him, then one of his English relatives on a different ship.

After seeing that all the Meriwether Y-chromosome profiles were the same, I began to look at the detailed results at each of the markers. It was clear that they all belonged to the clan of Oisin (pronounced *O'Sheen*), which I named after the mythical Irish hero but which is also known, much more prosaically, as haplogroup R1b. This is the most frequent clan in Britain and so I was very familiar with it. The clan links Celtic Britain with Iberia, from where Mesolithic maritime hunter-gatherers had moved north along the Atlantic seaboard to Brittany, in France, and on to western Britain and Ireland. However, this phase of ancestral migration took place at least six thousand years ago. Because this movement into Britain had been so early and so numerous, the descendants of these early Mesolithic Celts are the genetic bedrock of the whole of Britain, and this makes the clan of Oisin very numerous. So the detail of the Meriwether profile was very familiar as I read across the table of results. However, one result stood out and registered as unusual. The marker called 385 has two DNA segments and hence two separate results on the profile. In the clan of Oisin these are normally scored 11 and 14. But in the Meriwether Y chromosomes the first of these, called 385A, was 13 instead of 11.

I asked Richard to bring over my laptop, on which I had a lot of DNA data stored, and quickly looked up how many Oisin Y chromosomes had 11 at 385A and how many had 13. A quick scan of the last few months' accumulated results from Oxford Ancestors customers (I had not used this marker in my academic research) showed that out of 1,285 Oisin Y chromosomes, more than eleven hundred had scored 11 at 385A compared with only fifteen who scored 13. So the Meriwether Y chromosome is not only completely consistent with that particular spelling of the surname, but also extremely rare. This means, in my opinion, that any Meriwether who also shares this Y-chromosome profile is almost certain to be a relative, even if the paper trail is incomplete. "Margo" was visibly excited by this piece of reasoning and promised to redouble her efforts to get hold of the Virginia relatives of Meriwether Lewis.

I had been so gripped by "Margo"'s accounts of her ancestry that I almost forgot to ask her for a DNA sample. She agreed, chose her pseudonym, and then prepared herself. Unlike the usual kind of DNA test that requires only a discreet cheek swab, chromosome painting needs a lot more DNA and asks for 15 ml of saliva. That doesn't sound like much, but believe me it is a lot to summon up especially when put under pressure to perform. Having discovered that a sharp lemon candy stimulates the salivary glands, I had a supply at hand. Richard and I left "Margo" to it. Five minutes later we came back, capped the vial, mixed in a preservative, and the DNA sample was ready to send to the lab.

My next appointment was with "Atticus Finch," who occupies a senior position at the society. As a child he had met Gregory Peck and was delighted to be named after his character in the 1962 movie *To Kill a Mockingbird*, for which Peck won the Academy Award. We met in his office overlooking Newbury Street toward a skyline dominated by the mirror-glass monolith of Boston's tallest building, the John Hancock Tower. The tower commemorates, on a monumental scale, the memory of John Hancock, one of fifty-six signatories of the Declaration of Independence and the first governor of Massachusetts. In "Atticus Finch"'s office there was an altogether more intimate reminder of the great patriot—John Hancock's yellow wingback chair. Newly restored to its original condition, it was a handsome and stylish reminder of the comfortable life enjoyed by some citizens in eighteenth-century Boston. There was only one place for my photograph of "Atticus Finch" and that was in the famous chair. (Only afterward did he tell me that this had been the first time he had sat in it.)

Like all my volunteers from the society, "Atticus Finch" knew an enormous amount about his long ancestry in the New World. It was dominated by English and Dutch immigrants, most of whom had arrived in America in the mid- to late-1600s. Among them was Cornelis van Slyke, born near Utrecht in the Netherlands in 1604. He was a carpenter and stonemason who had come to the Dutch colonies of New Netherlands in America as a thirty-year-old. There he was indentured to Kiliaen van Rensselaer on his farm near Albany, New York, on a fixed-term contract that included repaying the cost of his transatlantic passage. Cornelis prospered and in 1664 obtained a deed to land at Schenectady, farther up the Hudson River. This was Mohawk territory and Cornelis became a trusted intermediary between the Dutch settlers and the Indians. His skill as an interpreter and a negotiator was evidently appreciated on both sides as the Mohawks also gave Cornelis some land at Cohoes, a few miles down the Hudson from Schenectady.

It was during the course of this work that he met and married Ots-Toch, the daughter of a full-blooded Mohawk woman and a French woodsman and trader named Jacques Hertel. Hertel had been born in Normandy, France, in 1603 and had arrived in America in the 1620s, to live among the Huron. This was a time when France and England were battling for the province of Quebec, and Hertel, resolutely anti-British, tried to persuade the Indians not to trade with them. When Quebec was returned to French rule in 1632, Hertel was recruited by Samuel de Champlain, the explorer and and eventual lieutenant general of New France, sent by Cardinal Richelieu to improve relations with the Indians.

"Atticus Finch" believed himself to be a descendant of Ots-Toch, and according to the chart that he laid on the desk in front of us, she was his ninth great-grandmother. When she married Cornelis van Slyke and settled in the Dutch colony, she became one of the first documented cases of an Indian woman who had left her tribe to live with Europeans.

Ots-Toch became very well known in her time and smoothed relations between the early colonists and the Indians. "Atticus Finch"'s line of descent came through Cornelis van Slyke and Ots-Toch's son, Jacques, and from there through six generations of Dutch and five generations of English Americans. His ancestry was a web of intersecting lines that went right back to the days before the Pilgrim fathers arrived in 1620— back to a time when Europeans and Indians were still feeling their way and 150 years before John Hancock put his signature, with a flourish, to the Declaration of Independence, sitting in the very chair from which "Atticus" had led me through this wonderful story.

The question on "Atticus"'s mind was whether he had inherited any of Ots-Toch's DNA. I knew the chances were slim. There were twelve generations between Ots-Toch and "Atticus"'s, and thirteen between him and her mother, the full-blooded Mohawk. With, on average, a two-fold dilution at every generation there certainly wasn't going to be much Indian DNA left in "Atticus." I worked out that we might expect one two-thousandth of his DNA to be from Ots-Toch. And even that figure, as we have covered, is subject to enormous variation. The two champions of genetic genealogy, mitochondrial and Y-chromosome DNA, would have shown a clear result as they are not diluted by DNA exchanges. But neither of them would work with "Atticus"'s genealogy because, first, Ots-Toch was a woman and so did not have a Y chromosome to pass on, and second her mitochondrial line would not have gotten past her son, Jacques. As a man, he would not have passed his mitochondrial DNA to his daughter, Lydia van Slyke, who was "Atticus"'s ancestor. So the only chance of finding a genetic link was through testing his entire genome and looking for any tiny remnants of Native American DNA that had survived in his ancestors. It was always going to be a long shot, but well worth a go. In no time "Atticus"'s DNA was on its way to the lab.

Our meeting with "Terry Malloy" was to take place outside Boston, so Richard and I hired a car and set off for Cape Cod. That is where "Terry" had been born and raised and where, by the sound of it, all of his ancestors had lived since they arrived in the seventeenth century. I wanted to see, if only very briefly, the replica of the *Mayflower* anchored near Plymouth and the reconstruction of the original settlement, the Plimouth Plantation, which retains its original spelling. Plymouth is on the way to Cape Cod, so we took a day away from the comfortable elegance of the society headquarters and headed off down Interstate 93. So many of the historic sites in Boston had a very familiar feeling about them because, of course, they were originally English and resembled similarly preserved examples back home. However, this illusion of architectural familiarity lasted only until we hit the highway. Once we had turned off to the little town of Plymouth, it was almost as if we were visiting a prosperous part of southern England we had not seen before. Almost, but not quite, as the gray-painted wooden houses were rather too elegant, and wooden houses in England are unusual. Even so, we were wrapped in the warm blanket of familiarity as we pulled up for a cup of coffee on the main street. Any sense of home was soon shattered by a nearby placard. It read "Ammo Special 7.62×39. $350 a case. Assault Weapons in Stock." We were not in a sleepy Devon town after all. We hurried past the *Mayflower* and the original Plymouth Rock beneath its protective cover, and on to Plimouth Plantation.

Despite my reservations about reconstructions I was impressed by Plimouth Plantation, where about thirty thatched wooden cottages lined two sides of a broad sandy street on a gentle slope heading toward the sea. Long tendrils of wild wisteria wound themselves around the green-leaved shrubs, and waving beds of cat-tail reeds lined the Eel River. Inside the cottages actors played the parts of the first settlers with immaculate conviction, never once betraying through language or action that they were anything but authentic colonists. I met the rather rakish Myles Standish, who told me he had come here from Holland, as he considered the Dutch too tolerant of Catholics. In another smoke-darkened interior Constance Snow was waiting for her husband to return from England, fearful that he may have been captured by French or Turkish pirates who preyed on the fragile transatlantic traffic supplying the colonies. In another, William Brewster explained how he he had been asked to conduct the daily services though he could not give the sacrament at Holy Communion as he was not ordained.

Far less convincing were the occupants of the reconstructed Wampmanoag homesite nearby. The bark-covered winter houses were just as interesting as the cottages but, by comparison to the enthusiastic role-playing by the English, the Indians, or "Native People" as a sign informed vistors they preferred to be known, were surly and reticent. The delicate relations between the two communities that I was to experience many times on my journey were already there in what was, after all, only a tourist destination, reinforced by notices asking visitors to avoid harmful stereotypes and not to ask any questions.

We pressed on toward Barnstable, where "Terry Malloy" and his wife had invited us for lunch. The woods grew thicker and the roads narrower. Enormous houses, mostly empty, stood back from the road. These, I later learned from "Terry," were the summer residences of the wealthy who spent the rest of the year in Florida. A few lawnmowers hummed as retainers kept the grounds in shape ready for the return of the summer visitors from the South, but overall it was silent, beautiful, and forlorn. The sun dappled through the trees, already golden-leaved, and every now and then we could glimpse the ocean. Inlets crowded with yachts sealed up for the winter came and went as we passed Falmouth, Sandwich, and finally Barnstable, all seaside towns in England but here disposed in unfamiliar juxtaposition. After losing our way several times we eventually found "Terry"'s house, in deep woods a mile or so off the main road. This was not one of the empty shells we had passed on our way but a neat, lived-in, one-story home. As soon as we got out of the car I could hear unfamiliar birdcalls, rich and strange, coming from the dense canopy of pines. A bright golden yellow bird, a kind of finch I think, landed on a branch close by then darted back into the undergrowth. In bleak comparison, our goldfinch back home has just one yellow feather on each wing.

"Terry," tall and lean with close-cropped gray hair, was waiting with his wife to greet us on the porch. Once inside, and seated comfortably in the living room, he began to tell me about Cape Cod, his ancestors, and himself. He was both laconic and content. "My life is not all that exciting," were his opening words in a slow, gravelly voice. "Oh, really," I thought to myself. He was retired now but had been an engineer, graduating from Brown University and specializing in electronic control systems. I was frankly dazzled by the projects he had worked on. Starting his career designing automatic pilots and instrument landing systems, he had graduated to control systems for supersonic wind tunnels, steel mills, submarine tankers, and radio-telescopes. "Submarine tankers?" I asked. "What are they?"

Pausing to collect his thoughts and take another sip of coffee, he recalled the project, now long abandoned, to load oil from the Alsakan fields at Prudhoe Bay onto huge submarines that would then travel under the ice to the North Atlantic and the east coast ports of America. It was to be, quite literally, a submarine Northwest Passage. The insuperable difficulty was not how to build these giants, but of transferring the oil, which had to be done underwater. To maintain depth the tanks had to be flooded with seawater as the oil was off-loaded, then flushed before taking on the next cargo. This simply could not be done without causing significant pollution, and, with the industry reeling from the Exxon Valdez disaster, the project was halted.

I could tell from the twinkle in his eye that "Terry"'s proudest achievement was his work on the radio telescope at Greenbank, Virginia. It had a diameter of 140 feet, yet had to be able to hold its position against the stars to within five seconds of arc as the earth rotated beneath it. That was the specification, but "Terry" built it to keep the alignment to within one second of arc routinely and to within one-tenth of an arc-second in good conditions.

Like many people, "Terry" began to be interested in genealogy after he retired. He wishes he had started earlier, before his father had died and taken so many family secrets with him. But fortunately the Sturgess Library in Barnstable had a fine genealogy section that gave him a good start. Then a chance meeting in Florida unearthed more history on Barnstable families, and as "Terry" worked his way through these, he found he was related to almost all of them. He had traced his patrilineal ancestors back to Edward, who arrived with four boys in West Barnstable in 1626. The next record was when one of the boys, another Edward, married Margaret Lombard in 1649. More ancestors peppered the intervening centuries, all of them from Cape Cod and mostly from the little corner near Barnstable. There was his great-great-grandfather Nathaniel, who had gone insane around 1800. His great-great-great-grandfather a generation earlier had been the captain of a coastal schooner that used to collect ice from Maine and take it to New York. He built a house on the water's edge so that, as "Terry" wryly suggested, he could land, start another baby, and be off to sea again on the next tide. It seemed to have worked, as his wife produced five daughters and five sons. "Terry"'s great-grandfather, a housebuilder, had put up the first house in Hyannis and also built the Wampanoag Meeting House in Mashpee, where he was later married. Every one of his ancestors had stayed within a few miles of where the first immigrant Edward had landed in 1626. The wide-open spaces of the West did not lure "Terry"'s ancestors to move away. They did not seek their fortunes in the California gold rush or in the money houses of New York City. They just stayed where they were.

"Terry" and his ancestors were completely enmeshed in Cape Cod, a point that was amplified when we drove the few miles to a local restaurant for lunch. "You see that house over there?" "Terry" said as we passed through a small township. "The one with the blue door? That belonged to my grandfather." Then, a mile or so farther on: "That house, the white one, was where I was brought up. It was built by my uncle, then my father built the smaller house next door, where he lived with my mother up until he died." As we wished them good-bye and drove back to Boston through the woods, I knew we had really touched the spirit of the place. Not extravagant, nor overambitious, just regular hardworking folk who loved their piece of America even more than their ancestors had done when they stepped ashore all those years ago.

We spent the rest of the week in Boston interviewing an eager succession of volunteers and taking DNA samples for chromosome painting. Many arrived with sheaves of documentation logging their ancestry back through the geometric progression of predecessors, doubling at every generation. Some, like "Atticus Finch," had been fascinated by genealogy from an early age, while others, like "Terry Malloy," had taken it up seriously only once they had retired. For some the passion went much further than mere enthusiasm. "Rose Sayer" told me she had spent fourteen hours a day, every day, for the past three years compiling her complete ancestry back seven generations, at which point the arithmetic meant she had 2^7 or 128 ancestors. She had tracked down 126 of these—and I mean "tracked down," with birth, marriage, and death certificates. As you can imagine, she was busily seeking the final pair to make her ancestry complete.

Quite a few of the society's members had already used genetics to overcome the "brick walls" that every genealogist comes up against at some point—those times when the crucial records cannot be found anywhere. By establishing a genetic link between two branches, usually through finding a Y-chromosome match, they could leapfrog the gap in the records and be confident that a connection existed where before they could only assume. By the end of the week I was left dazed by the sheer time and effort that the members had invested in their search for their roots. There was never a whiff of concern about enlisting DNA to assist them. Several members knew more than I did about the latest classification systems for particular branches of the Y-chromosome tree, which I thought a fine development.

All the volunteers so far had been European Americans, as I had intended in New England, but toward the end of the week a society staff member, "Harry Lime," asked if I could include a friend of his who lived in the same apartment house. His friend, who soon became "Virgil Tibbs," was an African American, and when we met in the society boardroom, he had a very different story to tell. "Virgil" worked as a history teacher in a school in Brookline, a well-to-do predominantly white town in Norfolk County on the outskirts of Boston. He had been to high school there as well, thanks to the busing program run by METCO, the Metropolitan Council for Educational Opportunity, which took him to school every day from his home in the inner city. I found our conversation over an hour and a half absolutely riveting, letting me glimpse a world I had never known. Whereas it was quite easy for me to imagine the lives of my other New England volunteers with their jobs in engineering or business, listening to "Virgil" as he very articulately described the everyday life of inner-city African Americans was absolutely gripping.

However, "Virgil" had not come to see me to give me a tutorial on the sociology of race relations in twenty-first-century Boston, but to explore a very specific family myth. He and his family had lived in the Boston area for a long as anyone could remember, and he had heard his grandmother say that she (and therefore "Virgil") was descended from King Philip. Fortunately I had read enough about early New England to know that King Philip was the son of Massasoit, the Wampanoag chief who had helped the Mayflower Pilgrims survive the winter of 1620 and negotiated a peace treaty with them in 1621. Massasoit lived for another forty years of largely peaceful alliance between the colonists and the Wampanoag. He sold land to Myles Standish and helped Roger Williams after he had been banished from the Massachusetts Colony on his way to found his own at Providence, Rhode Island. After he died in 1661, his sons Pometecomet, or Metacomet, and Wamsutta requested English names from the colonists. Metacomet was duly named Philip, and his elder brother Wamsutta became Alexander. Metacomet became the leader of the Wampanoag on his brother's death the following year.

Although he strove to continue the good relations with the colonists that his father had forged, increasing demands from the English finally turned Metacomet against the colonists, and he determined to stop any further expansion on their part. He attacked the Plimouth Plantation in 1675 but was decisively defeated by the colonists, and took refuge in swampland in Rhode Island. He was hunted down and shot the following year, his wife and child sold as slaves. In a final grisly reminder of the brutality of war, King Philip's head was mounted on a pikestaff at the entrance to the Plymouth Fort where it stayed for twenty years.

Five thousand Indians and 2,500 colonists were killed in the war. "Virgil" was hazy about his precise line of descent from Massasoit and King Philip, and although he did have some documentary evidence that his great-grandmother had received an annuity granted to descendants of Massasoit, it had been disputed by another family. "Virgil"'s cousin had a specific reason for wanting to establish his descent from King Philip, which was to support his application for membership of the Wampanoag Nation. The reason he gave was that he wanted the reinforcement of the feeling of connection to the place where he lives that an ancestor among the Wampanoag would give him.

As "Virgil" explained to me, black people don't feel that they have roots. They may be living in Harlem or Chicago or Tallahassee or San Francisco, but that is not really where they are from. Even in the South, that isn't really where they are from either. I asked him what he felt about African Americans who had reinforced their links to Africa, with or without a genetic test of the sort Rick Kittles and Gina Paige had developed. He told me that his grandmother had gone back to Senegal, where she thought her ancestors might have been from, but for "Virgil" the slave trade was a long time ago, "sort of hollow," and it was time to move on. "Everyone makes a big deal of race, but now we are all just Amazon-shopping, Big Mac–eating, Gap-clothed Americans." But to be a descendant of a Native American, a people who had been in this country for thousands of years before any European or African had set foot in the place—that was really something. Richard passed him a glass of water, "Virgil" filled the tube with saliva to the line, and the DNA that might hold the answer was soon on its way to the lab.

"Virgil" had been our last appointment in Boston. What a difference a week makes! What started as a rainy weekend facing a seemingly overwhelming and terrifying prospect of getting nowhere, ended in a slow walk back from Newbury Street to our hotel across Boston Common on a warm early autumn evening. The welcome we had received from "Atticus Finch" and the other members of the New England Historical Genealogy Society, the willingness of the volunteers to part with their DNA and to tell us about themselves and their families, could not have been surpassed. Richard and I sat down on a bench in the sun close to Swan Lake. He had sat in at every interview, worked the voice recorder, brought glasses of water to parched volunteers, and coaxed them to fill the sample tubes when the talking was done.

As we sat there after a very full week, I was contrasting the warmth of our reception in Boston to the time I traveled a hundred miles to give a lecture to a genealogy society in Northhampton, England. After I talked for an hour and answered questions for a further thirty minutes the organizers came round with a very welcome cup of coffee and a plate of cookies. I picked up a Rich Tea biscuit and my hand moved toward a Jammy Dodger on the other side of the plate. Before I could pick it up, the plate was whisked away. "Only one biscuit I'm afraid," came the stern admonishment.

I never again accepted a lecture invitation from an English genealogy society. On the bridge about a hundred yards away from where we were sitting, a saxophonist was playing. Low sunbeams illuminated hundreds of mayflies that were performing their mating dance, rising in columns, then falling back toward the water. Then something remarkable happened. The saxophonist let out a loud blast, and every one of the mayflies responded by flying upward much higher than before. It was as if they were dancing to the music.

13

Heading West

As the week in Boston passed and we became more and more immersed in old New England, I could not disguise my excitement and anticipation of the next stage of the journey. Our hotel was opposite the Tufts University Medical Center, about a mile from South Station. In the dead of night when the wailing sirens bringing emergencies to the hospital had quieted down, I caught the distant sound of a train. Quite unlike the frenetic adrenaline-bursting shrieks of the ambulances' "Get out of my way!" "Get out of my way!" the train sounded mournful, almost apologetic, a two-note harmony in a minor key: "Excuse me, but I must get through." "Excuse me, but I have to get to where I'm going."

When departure day came, our suitcases bulging with kind gifts of books and personal genealogies, we crammed everything into a taxi for the short ride to South Station. The large metropolitan stations in America are still quite wonderful, even though—or perhaps because—most of them were built for an era long gone, before planes broke their monopoly. Quite unlike the tearing hurry of airline terminals, the pace is slower

The California Zephyr crosses Colorado en route to Denver.

and, if you arrive with time to spare, relaxing. We settled down at a table on the stone concourse and ordered a couple of hot drinks. Tea for Richard, coffee for me, and then nuts and fruit for the journey ahead. We got out our train map and looked at the route. It seemed a very long way from Boston to Chicago, Denver, Salt Lake City, and finally Emeryville on San Francisco Bay. Even Chicago, which I had always thought of as about halfway across America, suddenly looked much nearer the East coast than the West. Which of course it is.

The low hum of the station was broken from time to time by a loud clacking sound, like a flock of noisy crows having an argument. It took some time for me to identify the source as the departure board resetting itself as the trains set off for their destinations. Our own train, the Lake Shore Limited from Boston to Chicago, moved up the board, and it was soon time to collect our luggage and climb aboard. We were not disappointed. The shining silver monster, two stories high, stood there hissing and steaming, waiting to transport us across America. We had booked a sleeping compartment luxuriating in its memorable description as a "Viewliner Roomette." It was on the lower of the two floors, about four feet across and eight feet long, with a window spanning its length. The "View" in "Viewliner" combined with the "ette" in "Roomette." The two comfortable seats faced each other across a table, thoughtfully decorated for chess. The seats converted into the lower of two beds, the upper one swinging down from the ceiling when it was needed. Given our weight difference and the possibility of collapse, Richard did not need much persuading to take the upper bunk. Our main bags checked, we stowed our hand luggage, leaving only a little room to move.

The train pulled away exactly on time and, with a blast of its whistle, wound its way at a leisurely pace past the maze of intersections, power lines, and industrial buildings, through the suburbs, and then out into open country. It was soon time for lunch in the café car on the top level. The train was wider than we are used to in England, and had roomy tables spread with starched linen cloths. The blue leatherette armchairs were at once experienced, tired, and extremely comfortable. The food was particularly good, cooked and served at a great pace by cheerful attendants. There are no separate tables on the Lakeshore Limited or any of the other long-distance Amtrak trains. Everybody shares, so in no time we were chatting away to our fellow passengers. Recognizing our foreign accents, they soon asked the first of many "Where are you guys from?" When asked to hazard a guess, most replied, "Australia."

As we moved away from Boston, the landscape changed almost imperceptibly from urban to rural. Fields gave way to woods. Many times we passed swampy ponds with the skeletons of dead trees reaching from the water like the fingers of drowning men. From time to time a lone tree, ablaze with red, showed off its autumn plumage among the restrained greens and yellows of its immediate neighbors. The countryside passed at a leisurely pace, around forty miles an hour by the feel of it. This is much slower than we were used to; our intercity trains regularly exceed a hundred miles per hour, and the Eurostar that connects London and Paris fairly flies along at almost double that speed. But neither the Lakeshore Limited nor any of the other trains we took ever exceeded sixty. I learned later that most of the track in the United States is not owned by Amtrak but by the freight companies whose mile-long trains, going in the opposite direction, we passed at regular intervals. I counted the cars on one of these—fifty, sixty, seventy—then I was distracted and gave up. There must have been more than a hundred. There is no need for them to go fast, and no need for the cargoes, such as coal, to have a comfortable ride, so there is no great incentive to maintain the track to the standard required for high-speed passenger travel.

Most of the passengers, we discovered, were on the train because they disliked flying. Jim, a middle-aged man from Los Angeles, was returning home after visiting his daughter in Boston. Unlike us he was not breaking his journey and after Chicago was going straight on. Three days and three nights on the train. That sounds like hell, but he was looking forward to it, though he did admit it got a bit tiring on the third day. As lunch drew to a close we arrived at Springfield, Massachusetts. The station was red brick and rusty metal—and deserted. No one got on and no one got off. Shortly after we left Springfield we crossed the first of many wide rivers that intersected our route across America. This one was the Connecticut. Soon we were back in the woods, not dense but dappled by the sun, and on our way to Albany, the capital of New York State. The woods looked as I imagined they had at the time of Ots-Toch, "Atticus Finch"'s Mohawk ancestor. It seemed ideal country for deer and other edible game.

Our driver was now applying the whistle with increasing vigor. At the approach to every crossing, no matter how small the road, the same melancholy blast rippled through the sleeping woods. Occasional homes interrupted the monotony of the trees, gray-slatted timber in plots with the abandoned swings of children long gone that parents didn't have the heart to dismantle. On one lawn an old rusting jalopy had collapsed at its final resting place, the grass carefully mowed around it as if this were a grave. These were most definitely homes, with all the paraphernalia of the living, not the pristine empty house-tombs of Cape Cod.

There were fields, but barely any of them were cultivated, and even those that were showed no signs of enthusiasm. In the distance blue hills appeared, the Catskills, I supposed from the map, but the trackside trees blocked any sustained view. The sun was low now. I closed my eyes, and blood red flashes flooded my retina as trees strobed the sunrays. The woods retreated, and we rolled into Albany across another wide river, this time the Hudson, and to the station where we were to wait for two hours to join up with the train coming from New York City. It was getting dark by the time we left Albany heading west toward Buffalo. The track passes along the shores of Lake Erie, but by then we were asleep. Richard and I are very used to sleeping on trains, as we go up and down to our house on the Isle of Skye, so we both slept well enough. Without knowing it we passed through Rochester, Buffalo, and Cleveland.

By dawn the country had changed again. Now we were in Ohio. The unbroken woods had gone and the land was much better cultivated. What trees there were grew in isolated clumps between dark green fields of corn. Gray-and-white cantilevered barns sat surrounded by woody hedges. A slow succession of small townships, their names displayed on water towers, passed by on each side. "Bryan (est 1840)"—I noticed that one, across the state line into Indiana, then Chesterton, where a Sunday market was in full swing.

As we got closer to Chicago, the land became more undulating, more woody by the mile. The vast cornfields thinned out and the housing changed from isolated farms to small apartment buildings surrounded by cars. Crumbling factories, some still breathing steam, lined the route. Getting closer and closer to the exurbia of Chicago, these gave way to chemical plants and refineries, burning off unwanted gas like giant candles. We crossed a canal, its water an opalescent blue that I have seen before only in the lakes of New Zealand. There it is caused by the diffractive effect of a fine suspension of glacier-ground rock. Here I am not sure the explanation is quite so natural.

The skyline of the city drew closer, and the trees all but disappeared. We passed street after street of boarded-up houses with For Sale signs nailed to the windows, the first hint of the subprime fiasco that rocked the banking system in America and has reverberated around the world. More small brick terraces of dreary uniformity as we passed the city limits. Some high-rises, and then we slid into Union Station. We had been on the train for twenty-two hours and covered 959 miles. Viewliner Roomette 3, car number 4920, had been our home, and we had not been bored for a second.

We were staying in Chicago for only two nights and had one very specific reason for being there and that was to talk with Rick Kittles, the scientific brains behind African Ancestry. We had arranged to meet on Monday so, it being Sunday, we wandered from our hotel in the theater district down the broad streets toward the shore of Lake Michigan. For the first time in a week we did not have appointments and could enjoy an afternoon of just being tourists. That evening we ate in the hotel bistro on the ground floor, where the de rigeur television was tuned, alternately, to baseball and American football.

Our visit coincided with the height of President Obama's fight to get his health-care reforms through Congress, and we were both taken aback by the political advertising, which we simply don't have back home. The messages were seductively presented but very direct. On the side of the reforms, we were told how so many Americans had no health insurance at all and that, even if you did but you lost your job and with it your insurance, you were in big trouble. The messages were interspersed with examples. Middle-aged couples who had their savings decimated when being laid off was followed by a serious illness. Young families whose claims were not honored by the insurance companies. On the opposite side, health-care-company campaigns emphasized the perceived advantages of being able to choose your own physician if you fell ill. We were vaguely aware of what was going on from news bulletins back home, and that our own National Health Service had been compared to a shrine of Stalinist ideology. Most surprising to my ears was the rhetorical question, "Do you want the government to run your health care?" inviting the answer, "No, certainly not." Back home that is one of the things we *do* expect the government to do as part of running the country. Even politicians from the Right know that if they were even suspected of wanting to dismantle the National Health System, where all treatment is free for those who need it, they would stand no chance of being elected.

Our hotel was art deco and very chic, but it also had one feature neither Richard nor I had ever seen before, which was revealed when a sudden movement on the wall caught my eye. The small mailbox in the hotel lobby had a glass-fronted chute running into it from the top and another coming out of the base. One floor down, on the ground floor, there was a large mailbox in the same position with the same chute coming from the floor above. What I had seen must have been a letter dropping down from an upper floor. This was too good to miss, and for a few minutes we were back as father and young son. Richard grabbed an envelope from the front desk and headed up to the top floor in the elevator. I went downstairs to the receiving box on the ground floor and waited. Sure enough, a few seconds later, the envelope flashed down the chute and into the box. Of course, schoolboys both once more, we did it again, but this time I was the one on the top floor and Richard was the ground-floor witness.

Brought up on the folklore of speakeasy Chicago still perpetuated in film, we kept our eyes open for evidence of the city's gangster past. We may not have found any, but we did eat out that evening in a restaurant that could, should—even *must*—have witnessed at least one gangland shooting. How else would we know about Chicago other than through the movies? Needless to say, the restaurant was Italian, and we were led into the dark interior where a lady—or could she have been a "moll"?—was riffing some jazz piano. The closely packed tables were crowded with diners, and all around the wall hung signed black-and-white photographs of film stars and theater actors. Many were strangers to us, but there was John Belushi, and over there Barry Humphries as his outrageous alter ego, Dame Edna Everage. As my eyes got used to the dark I scanned the other tables expecting to see, at the very least, George Raft as "Spats" Colombo from *Some Like It Hot*—that must be him in the far corner, his back to us, leaning over the table and talking to his fellow diners in a low voice. The piano gave way to a cabaret act. We went on eating. Spats and his party left quietly enough. We lingered over the cherry crisp and ice cream. The cabaret ceded to another pianist. There were no shootings, no police raids that night—but we left with the feeling that there just might have been.

Union Station, Chicago, is the hub of the American rail network. From there lines radiate east to Boston and New York and south to New Orleans. But the real excitement of the place stems from the lines heading west, of which there are three. The northern route, taken by the Empire Builder, pushes up through Wisconsin and Minnesota toward the Canadian border, then straight west through North Dakota and Montana, reaching the Pacific at Seattle. The Southwest Chief takes the southerly route through Missouri to Kansas City, Colorado, and Albuquerque, New Mexico. From there it crosses the desert to Flagstaff, Arizona, and eventually meets the ocean at Los Angeles. We had decided on the third, central route taken by the California Zephyr. Originally I had wanted to visit the Oglala Sioux reservation at Pine Ridge, South Dakota, site of the Wounded Knee massacre and more recently of the bloody standoff occupation by AIMs in the 1970s. Not "ancestry informative markers" on this occasion, but the American Indian Movement once led by the charismatic Russell Means. Having read his autobiography, *Where White Men Fear to Tread*, I definitely wanted to meet this man, and I set about trying to reach him. Though Pine Ridge is equidistant from the routes of both the Empire Builder and the Zephyr, the timetable forced a choice between renting a car in Fargo, North Dakota, at three in the morning or from Denver, Colorado, at seven. Having seen the eponymous Coen brothers film, and with winter on the way, we chose the Zephyr.

After our meeting with Rick Kittles (which I relate elsewhere), our luggage was growing as books, pamphlets, and reprints of scientific papers, all indispensable research material, were crammed into our three suitcases. Unable to find a Skycap, or a working elevator, at Union Station, we commandeered a reluctant trolley with an unhelpful determination to veer to the right as far as it could. We eventually arrived at the top of a flight of steps leading to the Amtrak reception area on the floor below. There I had a vivid moment of déjà vu. Even though this was my first visit to Union Station, the steps looked oddly familiar, with highly polished brass art deco handrails on either side. But where could I have seen them before? I had a clear vision of a child in a stroller hurtling downward, but try as I might, I could not place the scene or the movie. I thought it might have been Martin Scorsese's *Goodfellas*, but that was filmed in New York, I discovered later. Only recently, while watching the movie on TV, did I see the stairs again. There they were in the final shoot-out between Andy Garcia, as one of Eliot Ness's agents, and Al Capone's deadeye assassin, played by Billy Drago, in Brian De Palma's *The Untouchables*. Eureka.

The Zephyr waited on track 7, looking altogether taller, longer, and shinier than the Lakeshore Limited that had carried us from Boston to

Chicago. We settled into our Viewliner Roomette, with much the same layout as before, stowed our hand luggage, and got out the chess set to begin the second leg of the onboard tournament we had started at Albany. It was three games apiece as we pulled out of Union Station at two o'clock, dead on time, on our way to Denver, 1,038 miles down the track. The train rumbled through the by-now-familiar urban landscape of exhausted factories wheezing their last, demolition sites, empty lots, and occasional newly built low-rise apartment houses that surround the railway tracks in and out of most large cities, not just in America but everywhere. After half an hour the surroundings got a lot more elegant as we passed through Naperville—as I was told by a fellow passenger in the lunch car, one of the wealthiest cities in Illinois. Half an hour later we were going through Aurora, Illinois. This name was familiar but again I didn't know why until Richard got it. It was the setting for *Wayne's World*, the hilarious pastiche of Midwestern America, written by and starring Mike Myers, that Richard and I had watched back home a few months earlier.

Our lunch companions changed, and we found ourselves sharing a table with Ed, a man in his seventies wearing a U.S. Army hat embroidered with "Paratrooper Airborne Division." He told us he was a veteran of both the Korean and Vietnam Wars, and that he was retracing his first-ever rail trip when, aged sixteen, he had traveled from his home in North Carolina to Seattle to enlist as a paratrooper. Ed had eventually become a full colonel and gained a Ph.D. in military history, but he looked back on his career not with pride but with a deep sadness. He had witnessed slaughter on a grand scale, where life meant nothing. Hundreds of thousands had died in the wars he had fought, wars he felt the American army had never been allowed to win. It was all a long time ago—and no time at all. Like many older Americans we met on our travels, Ed told us he felt the country had lost all sense of pride in itself. We left him with still-sad eyes.

Outside a new greenish gold color appeared in the fields alongside the corn. This, I was told, was soya, a crop that, unlike corn, is not grown back home. The fields got bigger still, and the woods got smaller, and then we reached the river—the widest so far. We were crossing the Mississippi at Burlington, Iowa. Even with well over a thousand miles to go until it spills into the Gulf of Mexico near New Orleans, the Mississippi was already three hundred yards across, already muddy and sluggish. At that rate it must take weeks for the water to get to the Gulf. As we crossed the river we were also leaving Illinois for Iowa. Back out in the country we passed a pile of red-painted corn augers; the corn fields grew bigger and the farms more scattered. These were the fields of dreams, but there was no sign of "Shoeless Joe" Jackson among the waving stands of corn.

Suddenly the track got much worse, and the Zephyr slowed to about forty miles per hour, swaying from side to side. This was "lateral movement" according to the announcement over the public address system. Back in the roomette, our drinks spilled and the chesspieces fell to the floor. If it was going to be like this for the rest of the journey, then I had made a serious mistake in choosing the train to cross America. We recovered the pieces and resumed our game, which eventually ended in a stalemate.

Richard and I climbed back upstairs to the dining car, and this time we were sharing a table with a powerfully built man whom we soon knew as Jesus. He was going to Denver to collect a car for someone back in Wisconsin. Originally from Ciudad Juárez, just over the Mexican border from El Paso, Texas, Jesus had three children and at forty one was already a grandfather. He and his family had come to live in Wisconsin to escape the violence of his hometown and to give his children a more secure start in life. We never found out if anything more sinister lay behind his decision to leave the drug-fueled hell of what has been described as the most violent city in the world. And, having recently watched *No Country for Old Men*, another Coen brothers movie, we certainly didn't ask. We didn't want our door locks blown out by a cattle bolt.

The day ended gloriously as the sun broke out from below the clouds and suddenly bathed the fields with a brilliant yellow light. The train was heading straight into the setting sun. Three egrets stood out brilliant white against a dark plowed field.

By the time we reached the outskirts of Omaha the sky was completely black, but there was enough artificial lighting from the city to be reflected from the surface of the other giant river of the Midwest, the Missouri, as we crossed the state line into Nebraska. We settled down to sleep, and as I did so I noticed how the wail of the siren was very slightly different from that of our earlier train from Boston to Chicago. Whereas there was no mistaking the mournful minor key of Lakeshore Limited, the Zephyr's whistle was much more confident. Instead of "Excuse me, I have a long way to go," it was far more, "Here I come and you'd better get out of my way," as if the energy of the country were tilting to the west. Once in Nebraska, the line improved and the Zephyr speeded up, going almost fast. As on the train from Boston, the driver delighted in his generous use of the horn with three long blasts at every single crossing. I could imagine the occupants of the occasional trackside house being awakened at the same time every night, then turning over and going back to sleep after the Zephyr had passed, heading ever westward. At 3:00 a.m., as I looked out into the darkness, I saw an inland lighthouse that threw its beam for miles across the flat black countryside. What it was doing there, at least fifteen hundred miles from the sea in either direction, I never did discover. An hour later an orange glow reflected from the underbelly of distant clouds marked the position of a town, fifty, maybe a hundred miles away.

Dawn arrived, gray and raining, somewhere in eastern Colorado. We had spent almost the whole night, at a steady pace, getting from one side of Nebraska to the other. The country had changed, and the first sign of that was the silence. No aubade from the whistle warned people to get out of the way for there were no people, and no crossings. It was empty country, no fields, no farms, only gently undulating grassland that stretched, brown and waving in the half-light, to a thin horizon. Was that a herd of buffalo in the distance? Only when we got close did they morph back into dark brown cattle.

We went upstairs for an early breakfast and in doing so went through the sleeping car. At that time it had the atmosphere of a mixed geriatric ward. Seniors, who in any event made up the majority of passengers on the Zephyr, were slumped in their seats, sound asleep. Each of them had a name tag around his or her neck, large labels in plastic pockets fringed with green, each marked "America by Rail." William Jackson, or rather William "Bill" Jackson, and Frank Przewalski dozed opposite their wives. It all looked very uncomfortable, leaving me wondering whether they had known what they were in for when they booked. But seniors in America, I discovered time and again, are a hardy bunch, determined to wring what adventures they can from the years that remain.

As the sky grew lighter and the train took a gentle curve, I could see straight ahead for the first time in hundreds of miles. On the western horizon stood the solid wall of the Rocky Mountains, snowcapped and seemingly impenetrable. Before us twinkled the lights of Denver, our destination. Gradually buildings began to appear out of the half-light. What looked for all the world like a large temple and a hotel materialized out of nowhere at a town called Lowes. More and more buildings came and went, and we passed row upon row of new agricultural machinery stockpiled in yards by the track. Then, much more suddenly than in Boston or Chicago, we were barely through the suburbs before we pulled into yet another Union Station.

It was cold, about fifty degrees, but the awakened seniors were disembarking in their shirtsleeves. From Denver the usual route took the Zephyr through the Rockies, on the sector that made the line one of the most popular scenic routes in the world, and no doubt the highlight anticipated by the "America by Rail" passengers when they booked. However, urgent repairs to the track near Grand Junction had closed the line and forced a detour far to the north. We had already decided to leave the Zephyr at Denver and head for the Indian reservation at Pine Ridge in South Dakota, then loop around and pick it up again in Salt Lake City.

The only problem we faced was that I had utterly failed to get hold of Russell Means or anyone else from the reservation—but that didn't stop us heading for the airport to pick up a car.

After two unhurried days on the train, it was a shock to be brought right up to date by the unwelcome urban bustle of the car-rental counter. But what car to choose? Our imagination fired by footage of the grizzlies we would no doubt encounter on the road ahead, we decided we needed something large and solid. We were directed to that section of the lot, and there we definitely found large cars even by American standards. There was a silver monster called the Suburban—a joke, surely, since it would have flattened any child it came across playing ball on the streets of leafy outskirts that movies informed me to expect in all American suburbs. (You could have fitted the suburbs of a small American city into a Suburban and still have had room for the dog.) In the end we chose the most American-sounding of models, the Chevrolet 4×4. Richard at once wired the car radio to receive signals from his iPod—don't ask me how—and we were ready. We tried once more to reach someone from Pine Ridge without success, then headed north anyway. After all, this was supposed to be a road movie, wasn't it?

Our plan, such as it was, was to get to Rapid City, South Dakota, which is about an hour from the Pine Ridge Reservation, but as we headed into the empty prairies it seemed a very long way off, across most of Colorado. More or less our last glimpse of urban civilization was the enormous Budweiser plant at Fort Collins, a few miles north of Denver. Beyond that the grasslands stretched to the north and east, while the Rockies still crouched on the western horizon. This was our first encounter with big-sky country, and it lay under high gray clouds gently twisting and writhing like cigar smoke. Nodding steel donkeys, dotted around, pumped water from deep aquifers to the surface. A small clump of distant windmills on the top of a low hill caught the weak sunlight, turning into a bleached Golgotha. Low limestone bluffs interrupted the grassland from time to time, the lives of their entombed residents remembered in a Dinosaur Museum announced by the huge figure of a *Triceratops*, incongruous on the skyline. Snow fences to the west reminded us that this country was not just big but high as well, and that as a consequence the winters were severe.

We pulled into the small town of Wheatland for refueling and lunch, and to make a decision. The diner was warm and welcoming, and we slumped at a small table with a red gingham cloth. The heads of bighorn sheep, pronghorn antelope, and white-tailed deer looked down from the walls. This was hunting country, and judging by the cards on the bulletin board, bristling with taxidermists. We ordered, then contemplated our position. We were on our way to an Indian reservation with no guarantees that we would meet with anyone who could help us. And we also didn't really know what we wanted to do when we got there, anyway.

Although it broke our self-imposed road-movie rules, we did have a plan B. I had found a guide service operating out of Sheridan, Wyoming, that had advertised day trips to the sacred sites of the Cheyenne. I had called them before and been told that they definitely did not go to Pine Ridge, advising us to avoid that reservation altogether (they did not say why). But we decided to give Pine Ridge one last try. I called the number I had been given, and again there was no reply. So plan B was activated. I called Sheridan and heard that, thanks to a cancellation, the agency's top man was available for two days starting tomorrow. We paid up and hit the road. Richard scrolled through his playlist and, as we rejoined the freeway heading north for Sheridan, the Steppenwolf riff that opens "Born to Be Wild" blasted out of the speakers. This was more like it.

To the right lay the dry, broken grasslands that stretched all the way to the Mississippi. Large numbers of pronghorn antelope, camouflaged brown and white against the drying grass, raised their heads as we passed by. Occasional farms, one or two with stockades of domesticated buffalo, came and went as we ate up the miles. To the right the Rockies rose like a low curtain, snowcapped even in October, and still looking very hard to penetrate. At one point we crossed the line of the old Oregon Trail, over which convoys of settlers moved slowly west to their ultimate destination, still more than a thousand miles away. Though I knew very well that these settlers and thousands like them had obliterated the old Indian way of life, I could also admire their fortitude and determination. I wondered what it must have been like, toiling for weeks over mile upon mile of unbroken prairie, up and down the rough undulations in a covered wagon, camping by dried-up creeks. What a relief it must have been to see the mountains for the first time, not the end of the trek by any means but the end of the monotony.

14

The Great Spirit

The Havasupai case, and to a lesser extent the legal arguments surrounding the remains of Kennewick Man, have brought into the open the clash between science and tribal beliefs. It was the attempt to understand a little more about this clash that had brought Richard and me to the northern plains, one of the most thinly populated and remote parts of America.

Every culture has origin myths that do not always survive unscathed when scrutinized by rational examination. In two regions that I know well, the effect of science on belief has been rather mild in comparison with the chasm that has opened up in the United States. The indigenous people of Polynesia, with whom I worked for many years, believe in the original homeland of Havai'iki, although they are uncertain exactly where that is. Genetics has pointed to the islands of Indonesia as the likely location of Havai'iki and not, as many believed following Thor Heyerdahl's thrilling adventures aboard the raft *Kon-Tiki*, in the Americas. In Britain in the Middle Ages the predominant origin myth was that our ancestors had survived the Trojan War and our first king,

The Medicine Wheel, Bighorn Mountains, Wyoming.

Brutus, had vanquished a race of giants. He began a line of kings that included King Arthur with his Knights of the Round Table. This changed in later centuries to what is the predominant myth today—that the English are descended from Germanic Saxons and that today's Irish, Scots and Welsh are the descendants of indigenous Celts. Genetics shows that this is not the case, and that the genetic bedrock of the whole of Britain and Ireland is fundamentally Celtic overlaid with a thin topsoil of Saxons and Vikings, nowhere more than 20 percent. And no sign of Greek heroes or Trojans. But so far these discoveries have had very little material effect on belief—and have not resulted in any court cases.

Among American Indians, origin myths are very much alive, and though the details differ among tribes they share one fundamental similarity: that their ancestors lived where they now live. They have always been there and came from nowhere else, which is why they find clandestine research into their origins by outsiders so offensive. Spiritual affinity to the land is of utmost importance, and is easy to appreciate when you visit the lands they occupy. I do not pretend to be in touch with the Great Spirit, in whose name the Indians are not owners but guardians of the land and all life that depends on it. But I do have enough atavistic sense, handed down unseen from my own tribal ancestors, to feel the thrill of connection to the land in wild country. This is not the place to attempt in any way to summarize the rich tapestry of Native American origin myths—there are many better sources for that. But one myth, told to us by our Cheyenne guide, Serle Chapman, deserves a place here.

Richard and I drove out to Serle's place the following day. Tall and fair with carefully braided waist-length hair, he was evidently very accustomed to guiding European visitors like us around Indian country, which is what he does for a living. We set out at a brisk pace to Sacred Waters, or Lake De Smet, a shimmering blue lake about three miles long and a mile wide, not far from Interstate 90. Serle gestured with his hand to the island where a Cheyenne ancestor called Roman Nose had come on a vision quest at the age of fourteen and stayed for four days and four nights without food and water. Here also Ta-Sunco-Witko, or Crazy Horse, the Oglala Sioux war leader, came for guidance in 1875 when They Are Afraid of Her, his three-year-old daughter with Black Shawl, died of cholera. As Serle spoke, at a rapid rate, the elements of the landscape, the low hill to the east, the island, the water—all filled with characters from the Cheyenne past. I had to interrupt because I could see that otherwise we would be treated to a familiar recitation, very fascinating but not what I was really after. I explained briefly who I was and why we were there, and that I wanted to hear about the Ancient Time, the story of his ancestors. The Cheyenne, he told me, had not always lived there. They once lived in the east, near the Great Lakes, but had moved west to avoid defeat and enslavement by the Pawnee. They were again on the move after the Indian Wars and his great-great-grandfather Yellow Wolf had led part of the tribe south to Arkansas to become the Southern Cheyenne. Typically, there was nothing so prosaic as dates attached to Serle's recounted histories of his tribe.

The rest of the story had to wait until we reached our next destination, one never to be forgotten and the point when I first realized the dual nature of truth. We drove for two hours across the plains, past the town of Gillette, Wyoming, where off to the left the land was scarred by an immense opencast coal mine. They are dreadful sights anywhere, but the way the gigantic machines inched their way across the floor of the quarry, tearing the earth and scooping it into unsightly heaps for onward processing felt particularly loathsome that day. Soon we turned off the main road, then past a series of sculpted rock canyons that follow the course of the Belle Fourche River until, rounding a corner, we caught our first glimpse of Mato Tipila, Lakota Sioux for "Bear Lodge," but known to many as Devils Tower and to nearly everyone as the setting for the space-abduction scene from Steven Spielberg's epic *Close Encounters of the Third Kind*.

We skirted the main parking lot and walked up to a grassy mound at the edge of the ponderosa pine forest to the east. There we sat on a stone bench with the simple inscription, "In honor of all dads who introduced their children to the outdoors." Away to the east the sacred Black Hills of Dakota ruffled the far horizon. The sky was an unusually deep blue, and the faint waning half-moon was poised just above the summit of Mato Tipila. The giant rock rose up in front of us, and I asked my first stupid question: "How high is it?" "How high does it look?" replied Serle. This was beginning to sound like a psychotherapy session. Wisely I did not give an answer, but I did look it up afterward. The answer is 1,267 feet. Serle obviously did not think it needed a height, and I agree with him. But we can at least have a description. It is a high flat-topped mountain, almost circular in cross-section, with near-vertical sides. They are not smooth, but striated with long vertical cracks going from top to bottom. Serle began his explanation.

A long time ago, a great bear stole a man's wife and took her back to his lodge to live with him. The man tracked the bear down and crawled into the lodge, in the roots of a tree, and finding the bear asleep, woke his wife and led her away. They hurried back toward their own lodge, but the bear woke up, realized what had happened and, with a great roar, burst out in hot pursuit. Faced with certain death should they be caught, the pair called on the Great Spirit for help. He answered their prayers by lifting the ground beneath their feet and, with them safely on top, raised it beyond the reach of the bear. That did not stop him trying to reach them and he started to climb the sheer sides. Every time he climbed a little higher he slipped backward, and his claws dug into the rock and gouged out the furrows that are now such a feature of Bear Lodge.

The great bear eventually gave up, but that left the pair with the problem of getting down. In the end this was not necessary as the Great Spirit lifted them to the heavens, where they became the bright star cluster, the Pleiades, in the constellation of Taurus. As we looked at Bear Lodge, we all knew that this story was true. It was such an improbable rock, and the claw marks were in plain view. Around the summit, just visible without binoculars, turkey vultures circled in sinister silhouette. Before the rock was climbed, Serle told us, the only birds on Bear Lodge were eagles.

We walked back down the mound to the car and stopped long enough at the official parking lot to see a sign informing visitors that "The Tower is held sacred by many American Indians and highly regarded by other peoples," and asking that they respect the place. The audience here are the four hundred thousand visitors who come every year to see Devils Tower, first made a U.S. National Monument by President Theodore Roosevelt in 1906. What most of them see is the alternative truth, an igneous intrusion of gray porphyry, a rock rather like basalt, whose vertical furrows were caused by the hexagonal crystal structure of the rock as it very slowly cooled. Erosion removed the surrounding softer sandstones to reveal the stark igneous monolith. An obvious challenge to eager rock climbers, it was first scaled in 1893 with the help of a thousand wooden pegs hammered into cracks in the rock. Today hundreds of climbers make the attempt each summer, though they voluntarily avoid the month of June, when Indian ceremonies are held around the base. As we drove back toward Sheridan, I continued to absorb the revelation of dual truths. To Europeans, Devils Tower is one thing, to the Cheyenne it is another. It was pointless to say one was wrong and one was right. In their own way both were true.

The following day we headed in the other direction, west into the Bighorn Mountains. From Sheridan they didn't look all that high. Cloud Peak, the highest summit, had a little snow on it but looked nowhere near its real elevation of just over thirteen thousand feet, even considering that Sheridan is already four thousand feet above sea level. Serle took us to a steep-sided canyon leading into the heart of the mountains. The Tongue River Canyon had been used by Crazy Horse as a safe haven for his warriors and their horses as they were being pursued by the U.S. Seventh Cavalry under Gen. George Armstrong Custer in the weeks leading up to the Battle of Little Bighorn. The cream-colored rock of the canyon wall was sculpted by erosion into fantastic pinnacles three hundred feet above the track.

Picking a spot near the clear, fast river that flowed down the canyon, Serle kicked the rocks to dislodge any resident rattlesnakes, and then continued his stories of the Ancient Time. Although I thought he was getting accustomed to us and seemed to be enjoying our company, he nonetheless warned us that he was not telling everything and was introducing deliberate inaccuracies to avoid the story being stolen. I appreciated his candor, and rather liked the sound of that as a literary device. At first Serle elaborated on Maheo, the Great Spirit, and the four realms of the sky. At the highest was the spirit, which was masculine. At the other extreme was matter, the tangible substance of the earth, which was feminine. In the space between were one realm for masculine with female spirit and one for female with masculine spirit. (These intermediate realms had been largely forgotten on the reservations—which, we were told, explained the present-day intolerance of homosexuality, which would not have been there in former times.)

The past, Serle explained, was divided into four periods: Ancient Time, the Time of the Dog, the Time of the Buffalo, and now the Time of the Horse. Where had his people been in the Ancient Time? I inquired. Where had they come from? They had come from the earth, he replied— from the hissing, steaming vents that break through to the surface in Yellowstone. That was when Ancient Time began. Before that there were no humans. Just like Bear Lodge and Devils Tower, here again was a dual truth: For Serle and the Cheyenne there is no memory of an ancestral journey across the Bering Strait or the Atlantic or from anywhere else.

Other tribes have similar stories. The Navajo myth, for example, sees today's world as the fifth in a sequence, their ancestors having escaped from hostile earlier worlds through rents in the sky. Viewed from this fifth world, these are the very same fissures in the earth from which the ancestors of the Cheyenne once emerged. The ancestors of the Iroquois were the sky people, inhabiting *Karionake*, or "sky world", which floated among the stars. There lived Sky Woman, the ancestor of all humans. She fell from *Karionake* to Earth, which at the time was covered in water. She only survived thanks to a snapping turtle who offered her his shell as a refuge. Thus began the myth of Turtle Island, common to many Indian tribes. In time Sky Woman had a daughter, Lynx, whose children with earthly males eventually populated the whole world. These are all my ludicrously abbreviated summaries of a very rich creation mythology, doubtless including the subtle inaccuracies that honor the propriety of Serle's world. They all have their mystical qualities, but none is intrinsically any more strange than the Old Testament account of Adam and Eve and the Garden of Eden. But while many Christians are content to regard that as a parable rather than a literal truth, the same is not the case for many Native Americans, as we shall explore a little later.

After we left the river canyon, we headed farther into the Bighorn Mountains toward one of the most sacred Cheyenne sites, the Medicine Wheel. Up and up the hairpin highway, higher and higher as Sheridan shrank from view into the haze of the prairie. Then the gradient receded, and we were driving through dense pine forest. On and on, past occasional clearings where white-tailed deer looked up cautiously as we drove by. After several miles the woods thinned, and we passed the tree line. Scrub willows grew alongside hidden streams, but as we climbed higher still even they disappeared, and we were on open grassland. A large flock of sheep appeared by the side of the road, and two of them broke away and headed toward us. Except they were not sheep but large white dogs, guarding the flock from bears and coyotes. There was not a shepherd in sight.

A few miles farther we turned off the road and drove along a rough track, then parked and got out of the car. We were very high up by now. The sun was going down over Yellowstone a hundred miles to the west, taking with it the last warming rays. It was getting rapidly colder, a consequence of the high altitude. As soon as we began the shallow climb toward the Medicine Wheel, I could feel the thin air. My lungs expanded, but I wasn't inhaling anything. The track led up through spare trees and rock-strewn meadows onto a wide ridge that, like the approach to some ancient Mayan temple, led up to a high point on the horizon. Thousands of feet below to the north, the Bighorn River sparkled in the low sunlight. The gaunt shadows of the few remaining trees grew longer. I was beginning to feel the strain of the climb in the thin air and stopped many times to catch my breath. And then, quite suddenly, we had arrived.

On every side the mountains sloped gently away, and there, on the plateau in front of us, was the Medicine Wheel, an assemblage of flat rocks arranged like spokes around a central hub. The perimeter was defined by stout wooden posts, and on the connecting wires hung hundreds of objects. What they were I did not know. Serle was silent. He stood well back. He did not speak, but he had already explained on the way up that, viewed from the center, particular stones on the circumference marked the rising points of the principal stars: Sirius, the brightest, the old woman, spirit of the earth. Rigel, the blue star in Orion, the blond girl. Red Aldebaran, in the constellation of Taurus, the old man, the red wolf. The twenty-eight spokes were the ribs of a buffalo, and the days of the lunar cycle, and of menstruation. Below the central hub was a cave, a sanctuary of deep earth, its opening where the divine breath reached the sky. Richard and I walked slowly round the perimeter. Among the tokens tied to the fence wires were pouches that, we learned later, were filled with tobacco as offerings to the Great Spirit. We did not think we should take photographs, such was the effect of the place, along with Serle's silent waiting. We rejoined him, then he approached the wheel without us, and left his own pouch. We walked back to the car in silence. By then it was getting dark, and the brightest stars were showing against the deep blue of the sky. We returned the way we had come. Serle headed west to Yellowstone to look for grizzlies. We never saw him again.

Because of our time with Serle and the appreciation, if only rudimentary, of the spiritual life of the Cheyenne, of the Ancient Time, of Maheo, the Great Spirit, of Bear Lodge, I began to feel uneasy about our scheduled visit to the reservation, where, our other prospective guide had warned us, we would see how the Indians lived now in poverty and squalor. This may or may not have been true, but I wanted my last memory of the Cheyenne to be on the high-mountain plateau under the stars and imagining the ancient ceremony of Massom.

We left Sheridan and headed north up the I 90 toward Billings, Montana, which was to be our overnight stop before Yellowstone. After a while we entered the Crow reservation and immediately the fields, which had up to then been well fenced and with good grass, got considerably less good. The farm buildings disappeared and were replaced by shabby townships of mobile homes and precarious shacks. We had been promised a day on the nearby Cheyenne reservation and a meeting with Serle's uncle, the keeper of the sacred arrows. However, the prospect of witnessing the modern-day poverty our current guide promised, and the depths to which the once-proud nation had fallen, threatened to divert our experience, and the book, away from its original intention. We rendezvoused at the Cheyenne Trading Lodge, just off the freeway, and very close to the Little Bighorn battle site. The lodge was a pastiche of faux-Indian artifacts run by a white guy. The casino nearby was seedy, run down, and under threat of closure. After lunch I took Richard aside and told him how I felt. Not for the first time, he felt exactly the same. We explained to our guide, who was fine about it, paid her for the day, and said good-bye.

The nearby government-run site of Little Bighorn was, we discovered, very well laid out and informative. It marks the place where an alliance of Cheyenne, Lakota Sioux, and Arapaho led by Sitting Bull and Crazy Horse defeated the Seventh Cavalry under General Custer. Little Bighorn, in late June 1876, was the only significant battle victory for Indian resistance among a long catalog of defeats. We saw, as every visitor does, the rows and rows of memorials of those who fell in the Indian Wars, and the memorials to Custer's men of the Seventh Cavalry, who were buried where they fell on the side of a low mound. As so often at sites like this, I felt I had learned something, but I could not fully appreciate its significance. I could not feel the noise and the fear as Custer's men, hopelessly outnumbered, sheltered behind the bodies of their fallen horses and emptied their rifles into the advancing Indians until the ammunition ran out. Their deaths after that were inevitable, swift, and merciless. News of the defeat reached Washington on July 4, Independence Day, in its centennial year. It was received with general consternation by a public accustomed only to news of victories. The battle is widely thought to mark the beginning of the end of the Indian Wars. The U.S. Army redoubled its efforts to defeat the few remaining tribes that had resisted expansion from the East. Realizing that there would be retaliation, the victors of Little Bighorn dispersed. Sitting Bull crossed the border into Canada, while Crazy Horse surrendered in 1877 and was fatally injured by a military guard in Fort Robinson, Nebraska. The final encounter in the Indian Wars was the Wounded Knee Massacre on the Lakota Sioux Reservation at Pine Ridge, South Dakota, in late December 1890, when the U.S. Seventh Cavalry killed 150 mostly unarmed Indians.

It had been a hot day at Little Bighorn, and as we headed into Montana toward Yellowstone, the car thermometer registered ninety-two degrees. (The temperature swings around here at this time of year are remarkable: The following week the whole area was covered by the first snow of winter.) We entered Yellowstone through the arch built to commemorate its establishment as the first national park by Teddy Roosevelt in 1908, and headed for Mammoth Springs. We had booked into the Yellowstone Lake Hotel, but a forest fire was threatening to put it out of bounds to guests. Mammoth Springs is one of dozens of thermal vents in Yellowstone and had that smell of rotten eggs that signals the presence of hydrogen sulfide in the atmosphere. Steam and boiling water hissed from dark fissures lined with yellow crystals of sulfur. Rather like the Bear Lodge/Devils Tower dichotomy, we were at the exit of the womb of the earth mother, the very birthplace of humanity for the Cheyenne, and at the same time at a geological location where the earth's crust is both thin and perforated. In Yellowstone we saw what most people come there to see, chocolate brown bison warming themselves by the steam from thermal springs, places where bears had torn bark from the trees, distant views of elk and a solitary coyote, but no actual bears or wolves.

We did not get to the geysers as the forest fire was spreading fast, and the roads had been closed. As guests with reservations we were, however, allowed access to the Yellowstone Lake Hotel, an elegant wooden structure built in 1891 and recently renovated in the style of the 1920s. From the lakeside jetty we watched dense plumes of smoke drifting across the water, colored pink and orange by the setting sun in an otherwise azure sky. Helicopters ran shuttles from the lake to the seat of the fire in a desperate attempt to confine it. Forest fires are part of the natural sequence of life and death in Yellowstone, and the pine seeds need a good roasting in the flames before they will germinate. This fire, code-named Arnica, was receiving so much attention because it threatened the hotel. It did not feel dangerous, but sprinklers were set up to dampen the vegetation beside the access roads with water, which froze at night as the temperature plummeted. Each evening the progress of the fire was mapped on a board, new arrivals canceled, and the hotel gradually emptied. The Arnica fire had been started by a lightning strike on September 13. Sixteen days later, when we arrived, it had spread to more than eight thousand acres. The following day another thousand acres had gone. During the day we could see only the smoke as it drifted across the lake, but at night flames leaped high into the sky. The next day the fire had crossed and closed the road to the south, which was our planned route out of Yellowstone to Salt Lake City. The hotel became emptier and emptier, but there was a good spirit and the lobby pianist kept on playing.

At four in the morning the fire alarm went off, and we all assembled outside ready to evacuate. There was no panic, but by now we really could smell the smoke as a change in the wind set the fire on a collision course with the hotel. Due to leave that morning anyway, Richard and I hastily packed our bags and headed off early. The route change meant we had to leave the park by the only exit still open and retrace our steps into Montana, then loop around to Butte, and get to Salt Lake City by driving through the whole of Idaho from top to bottom. The road took us over the continental watershed, where rain falling now headed west to the Snake River and the Pacific, rather than back along the Yellowstone River to join the Missouri on its long journey to the Gulf of Mexico. Are there any raindrops, I wondered, whose destiny is shared, some water molecules traveling in one direction and some the other? I think there must be.

We eventually covered 636 miles, the longest distance I have ever driven in a single day. We knew we had to get to Salt Lake City to pick up the Zephyr for the final leg of our westward journey. The Utah capital was full of preteens going to a Hannah Montana concert. Merely a low shed near the bus station, the Amtrak station in Salt Lake City couldn't hold a candle to Chicago or Denver, and had a very temporary feel to it. The long diversion caused by track repairs meant that the train was due to arrive at a barbaric 2:15 a.m. rather than a more civilized 11:00 p.m. Arriving in Salt Lake City, the headquarters of the Mormon Church, I had already had my expectation of a haven of calm and contemplation dented by the roadside signboards on the way in. None more so than "Call Laser Lipo 'We suck . . . FAT.'" I had also imagined that it was completely safe, and it probably was, but the darkened sidewalks set off all the familiar urban alarm bells and we found temporary refuge in a hotel until the train arrived. The Zephyr eventually came in, and we located our third and final Viewliner Roomette and went straight to sleep. I would have liked to see the Bonneville Salt Flats, scene of many attempts at the land-speed record, and what is left of the Great Salt Lake, but I was already miles away.

When I awoke I was suddenly struck by an all-consuming sadness. This was our last day on the train, and almost the final day of our three weeks together. Richard was leaving for England to begin college, and I knew things would never be the same again. He was eager to begin his new phase of life. He had said, a day or so earlier, that two weeks is long enough in the company of parents and that he needed to be with young people. He was right; it was right; it was what should happen. As I looked at him asleep in the upper bunk, slow tears formed in my eyes.

We were definitely in desert country now. Sagebrush and yellow-flowered ragweed flanked the track. Lonely cattle, Herefords, possibly, to my British way of thinking, grazed on nonexistent grass. How did they manage out here, these beasts of the apple-flanked meadows of England?

This was a parched and thirsty land. Then, all of a sudden, we crossed a proper river about twenty yards across, flowing east away from the sea and into the desert. The drops that made this river would never reach the sea.

There seemed no end to the Great Basin, but eventually we reached Reno, Nevada, at its western fringe. Reno is nowhere near as spectacular as Las Vegas, but still crammed with casinos. A few miles farther on we crossed the last of the twelve state lines of our trip. We had traveled more than three thousand miles and there were now only just about two hundred to go to the Pacific. But we were still 4,500 feet above sea level, and the necessary descent began soon after we crossed into California. We slid slowly down the most difficult section of the track between the hills of the Sierra Nevada covered in redwood and Douglas fir, sparse at first then in dense stands on either side. The line had been built in the 1860s by ten thousand imported Chinese laborers, many of them the ancestors of the present-day residents of San Francisco. The numerous tunnels, embankments, and cuttings blasted from the rock had all been made by manual effort long before the age of the machine. Once again I found myself contemplating the dual nature of events. On the one hand the railroad was an absolutely magnificent engineering achievement, built with determination, skill, and courage. On the other it was the steel needle that had been thrust through the thin skin of the prairies, injecting its paralyzing anesthetic into the dying world of the Indian.

My ears began to hurt. After more than a week at elevations well above five thousand feet, the sudden descent was painful as my ears equilibrated with the higher air pressure. Eventually the gradient flattened out, and we were in California's Central Valley. We saw our first palm tree at Rockland and our first orange tree a little farther on. And another thing we had almost forgotten—soil. Here was rich farmland, black under the plow. The train speeded up, and it was then that I noticed another change in the tone of the whistle. No longer the assertive "Out of my way" of the Midwest prairies, its tone now was much more exuberant, more triumphant. Now the Zephyr seemed to be saying, "Look at me. Look how far I've come!" Through Sacramento, the state capital, and on toward the ocean, towns were dotted all over the flat and fertile plain. We met the sea at Grizzly Bay, where dozens of mothballed navy ships were anchored offshore awaiting demolition, and edged our way along the eastern edge of San Francisco Bay toward our final destination at Emeryville. The train and the track were diminished by the urban surroundings. No one took any notice of the Zephyr—no longer the only man-made thing for miles, the pioneer of the desert and the high sierra. After such an epic journey surely the tracks would be lined with spectators, cheering and throwing their hats into the air. But no, it was just another busy day with no time to stand and stare. And then we arrived, not in the marble art-deco splendor of Chicago's Union Station, but into the uncovered platform and low-level booking hall of Emeryville, California. This was as far as the Zephyr went—the track does not cross the bay into San Francisco itself. Apart from the section closed for repairs, we had traveled the whole 3,497 miles "from sea to shining sea."

We spent our last night together, without ceremony, at the airport hotel. And we finished our chess tournament. Richard won by eight games to seven. The next day we packed his bag and took the shuttle to the airport. We waved goodbye, father and son, and I watched as Richard moved up the security line. One last wave as his bags went through the X-ray machine, and he was gone.

Back at the hotel I sat at the water's edge. Pelicans and terns flew over the open sea, waders picked at the mudflats as the tide receded. The runway was about a mile away across the water, and I watched the comings and goings of the planes. After what seemed like hours the giant 747 I knew was Richard's taxied to the very end of the runway, turned, then paused. Was something the matter? Would the plane be recalled to the gate? I almost hoped so. But of course it wasn't. The roar of the engines throbbed across the bay, and the silver bird began to move down the runway. Slowly at first, then faster but never fast enough—or so it always seems with a 747. It passed the control tower, then the vast hangars, then turned its nose slightly upward to greet the sky, and left the ground. The sky was a brilliant blue, and I followed the plane as it climbed very gently straight ahead. It gradually diminished into a small white bird. How can anything so large look so tiny? The white bird got smaller and smaller until I needed my binoculars to see it. Eventually it turned to the east. I still followed as it became a tiny speck, then put the binoculars down to clear a piece of dust from my eye. When I tried to find the plane again—like Richard's long childhood—it had vanished.

15

The Persuaders

The great bird that had taken Richard back to England returned the next day with another special passenger. While Richard had been my traveling companion and research assistant on the first leg of my journey around America, I would be accompanied by Ulla on the return leg, from west to east. Once more sitting by the edge of the lagoon outside the hotel, and again in brilliant sunshine, I scanned the southern sky for the red-tipped wings of the Virgin Atlantic 747 that was bringing her from London. Minutes before the scheduled arrival time, I picked out the landing lights shimmering in the afternoon haze across the bay. Slowly, slowly the plane swung around and flew low over the San Mateo Bridge, descending all the time until, with a puff of blue smoke, it was on the ground.

I was in San Francisco primarily to meet again with the scientists from 23andMe, the genetics company who had been doing the chromosome paintings. I wanted to replenish my stock of DNA-sampling kits and, I hoped, get together with some of their customers who had already had their portraits painted. That evening, as Ulla and I sat in the

Golden Gate Bridge, San Francisco, California.

glass-fronted lobby of our hotel, overlooking the bay, we noticed a blimp cruising in the distance. It looked perilously close to the airport, and I wondered if it had lost its way, or its controls. Then I noticed the genetics company logo on the side. Things must be going well.

We soon got talking to our fellow guests. These were brief encounters, usually opened by a distinctly American "Hi there, where are you guys from?" These remarks were invariably directed at Ulla, who is tall and blond in a typically Scandinavian way. By the time she had explained that she was from Jutland, on the west coast of Denmark, but had been living in England for many years, nobody was the least interested in my origins. Even so, I did get the chance to join in the conversation when Ulla had explained why we were there and what we were planning to do. Everyone was very interested in what I told them about the project, and one or two had even read *The Seven Daughters of Eve*. It slowly dawned on me as the evening drew on that we had stumbled across the perfect DNA-sampling location. A comfortable open space dotted with armchairs, where conversations were easy to initiate and nobody felt threatened. There were secluded booths for the sampling itself and visitors checking in from all over America. It was perfect. That was how, that very first evening, we met the woman who chose to become "Sugar Kane," at least for the purposes of our genetic research.

"Sugar" was visiting one of her children, who was a nurse at San Francisco General Hospital. She and her husband had flown all the way from Jacksonville, Florida. "Sugar" was a retired college secretary who had an interest in her family history, not perhaps as intense as I had witnessed in Boston, but enough to wonder about where her ancestors were from, particularly her grandmother's origins. She had been only four when her grandmother died, so the memories of her were sketchy at best. However, when later in life she had been leafing through some old photo albums, she realized for the first time that her grandmother stood out from all her other relatives by being much darker skinned and with long straight dark hair and high cheekbones. To "Sugar" she looked as though she could have been Native American. Unable to verify this through any genealogical records, she nonetheless felt drawn to discovering more about Native American spirit life, and began to attend classes run by a medicine woman called Pansy Hawk Wing. She was so taken with this experience that she and five other women, with their husbands, traveled to Pine Ridge Reservation in South Dakota with Pansy Hawk Wing to embark on a vision quest. She and the other women were confined to a special lodge on a hilltop, where they were left overnight without water or food. "Sugar" had her own drum, called Ladybug, on which she drummed through the night. In the morning she and the other women walked downhill to meet with Pansy and the men. From there the women and their husbands entered the sweat lodge, heated, dark, and airless. After some hours they emerged and the women crossed to the moon lodge, where they were "pampered"—her own word—by other women. "Sugar" had found the whole experience extremely invigorating.

I was not sure what to think. It was certainly not easy to imagine "Sugar," who looked so typically European, with her blue eyes and dark blond hair, in the sweat lodge at Pine Ridge. I subsequently discovered that sweat-lodge "experiences" were widespread, even outside America. (There was actually one advertised near London.) There is always the chance of cynical exploitation, but there was no doubt at all that "Sugar" had gained tremendous pleasure and self-awareness during her week at Pine Ridge, part of a continuing journey that began when she saw the photograph of her grandmother. It would have been worth it even if "Sugar" had no Native American ancestry, but you will not be surprised to hear that she did not need much persuading to take part in the project to check out her DNA.

As chance would have it, the more dangerous side of the Native American spiritual experience emerged only days after I met "Sugar" and her husband. Three people died and eighteen were hospitalized after a sweat lodge in Angel Valley Retreat Center in Sedona, Arizona, went badly wrong. James Arthur Ray, the organizer, who ran the "Spiritual Warrior" event of which the sweat lodge was a part, was arrested and charged with manslaughter. At his trial in 2011, he was eventually cleared of manslaughter but convicted on three counts of negligent homicide.

One very good reason to be in San Francisco was to have a longer conversation with Joanna Mountain than had been possible on my first visit to 23andMe nine months earlier. Ulla and I steeled ourselves for the short journey down Highway 101 to the company headquarters in Mountain View, joining the headlong rush of cars hurtling south toward San Jose. On this occasion we got off at the correct exit to the relative calm of the wooded avenues of Palo Alto.

Mountain and I started by comparing notes on something that was proving to be one of the most interesting aspects of the research, and that was the wide range of attitudes toward genetic testing held by different groups. Talking to our Cheyenne guide, Serle Chapman, in Wyoming, I had already sensed the intrusion that his people had experienced from mainly white scientists who, they felt, were trampling on their own origin myths. Joanna confirmed that she had seen the same reaction, epitomized by the Havasupai case, when researchers had gone beyond the original study designs by allowing samples to be used in ancestry projects for which there was no consent. This had created a backlash among Native Americans against genetics generally that was only now beginning to be reversed by careful dialogue. However, while there was no appetite for genetic ancestry information among Native Americans themselves, there was a great hunger among other groups to prove, through genetics, at least some Native American ancestry. It was not only European Americans like "Sugar Kane" who were interested in looking for a genetic connection to an indigenous ancestry: I had seen a similar desire in "Virgil Tibbs," my African American volunteer from Boston.

Mountain also had her own fascinating story to tell, not about herself but about her husband Heyward Robinson, who is from an old South Carolina family. Understandably Heyward had his DNA analyzed through his wife's company and found, perhaps not too surprisingly, that his chromosome portrait showed no trace of orange, the sign of some Asian or Native American ancestry. However, a subsidiary test on his mitochondrial DNA came back with something quite different. The results showed without a shadow of doubt that his direct maternal ancestor was descended from Aiyana, not a European clan mother at all but one of the five matrilineal ancestors of Native Americans.

Heyward Robinson's mother was an avid genealogist who had traced the family back to the seventeenth century with no sign of anything but European ancestry. On his last trip back east to see her, Robinson had visited the town his matrilineal ancestors were from and, helped by another genealogist friend, found an ancestor seven or eight generations back who seemed to have come from nowhere. Heyward discovered that his fifth great-grandmother, Elizabeth Keys, was a resident of Orangeburg, South Carolina. Her father, John Keys, had been a fur trader, but the identity of Elizabeth's mother was a complete mystery, despite many years of research by family members. Mountain checked out more of her husband's matrilineal relatives and found the same Aiyana mitochondrial DNA in all of them. She also found a small "orange" segment of Native American chromosomal DNA in his mother and sister. This was consistent with a Native American ancestor whose DNA had survived to his mother's generation and had been passed on to his sister, but which, through the lottery of genetic transmission, had not gotten through to Robinson. Judging by the one remaining "Native American" segment in his mother's chromosome portrait, it had joined the family tree around three hundred years ago. This was at a time in the seventeenth century when people were brought over from Switzerland to act as a buffer between the English colony in Charleston, South Carolina, and indigenous groups in the surrounding territory. In those ideal conditions for intermarriage, it is not unrealistic to suppose that Heyward's mysterious ancestor was the fruit of such a union.

What I found particularly revealing about this story was not just its illustration of the sublime persistence of mitochondrial DNA outlasting all the chromosomal DNA, at least in Robinson's case, but also what

Mountain told me about his reaction. He was both thrilled and fascinated, and it shifted his view of himself. One evening the family was
watching Terrence Malick's *The New World*, a movie which depicts the
early European settlement of America and features the heroism of Poca-
hontas in saving the English colonist John Smith from execution. To
Mountain's great surprise, her husband turned to their two children and
said, "Kids, you have deep roots here." For Joanna, witnessing this unexpected cultural outcome was, in her own words, "something else."

Another reason to visit 23andMe once again was to speak with Mike
MacPherson, its statistician. I wanted to check some of the finer technical points of chromosome painting, but soon after he joined us, the
conversation turned to the question of the interpretation of the chromosome portraits of Native Americans. Every indigenous North American sample ever analyzed shows a combination of orange and blue in its
portrait. Did that mean Native American DNA was always a mixture of
Asian and European, and, if so, had that mixture arisen within the last
five hundred years, or was it already there in the first Americans? As we
saw in chapter 11, the reference populations for the orange "Asian" segments were from China and Japan rather than from northeastern Asia,
the more likely source for those ancestors who arrived from Siberia. Certainly mitochondrial DNA evidence had shown that most Siberians had
arrived there from central Asia rather than China. Central Asia is getting pretty close to Europe, so the question of the boundary populations
starts to come into it.

When MacPherson had been developing the chromosome-painting algorithms, he had certainly considered what would happen at the
boundaries between the three continents. When the world is crudely
divided into just three zones—Africa, Asia, and Europe—for the purpose of the chromosome portraits, the situation is bound to be complicated where those zones meet. Anticipating this, MacPherson got hold
of DNA from Uzbekistan, in central Asia but much nearer Europe than
China and Japan, and from the Berbers of Morocco, where Europe meets
Africa. Sure enough, these showed a complex mixture of segments origi-

nating from the continental populations on either side of the boundary. The portraits look as though they have come from particularly confused populations with ancestors from both sides of the line separating the continents, but this is just an artifact of the way in which the source populations are defined. It would be the same wherever you drew the lines, because that is what ancestors do: They move about and cross lines. When MacPherson tried this out in Europe, trying to discriminate between chromosome segments with genetic origins in the north and the south, everybody was a mixture of the two. It was one of those instances when more detail confuses rather than clarifies the picture. Much better to accept the inaccuracies of the crude continental divisions and bear them in mind when sitting back and gazing at the chromosome portraits.

I asked MacPherson the obvious question: If you could go back five hundred years, before recent European contact, what would the chromosome paintings of the indigenous Americans have looked like? Would they be completely orange, in which case any blue would indicate some recent European ancestry, or would they be a mixture of orange and blue, reflecting the ultimately central Asian origin of Siberians and the ancestors of the first Americans? Somehow I wasn't surprised when he told me he had already tried to simulate this. His answer was that five hundred years ago the indigenous Americans would have had two-tone portraits of 75 percent orange, plus or minus 10 percent. The rest would have been blue. His figures meant that in individual contemporary Native Americans, any mixture with figures well outside this range was probably the outcome of a mixed ancestry.

I had always assumed that any incoming ancestry would be European, but then I remembered something Serle Chapman had told me as we sat looking at Bear Lodge in Wyoming. He recounted the tale of an Indian raid on a railway construction site in the Rockies during the mid-nineteenth century. Since thousands of Chinese had come to help build the western sections of the railway, it was a mixed labor force of white European Americans, many with an Irish background, and Chinese that came under attack. After a brief exchange of fire, the heavily outnumbered Europeans ran off, but the Chinese stayed where they were. To the Indians they were not the enemy, and also they looked rather similar due to their shared Asian ancestry. So they joined the Cheyenne and no doubt in time passed segments of their completely Asian chromosomes on to future generations, thereby increasing the overall orange component of Cheyenne DNA. Finally I asked Mountain if I might talk to some of her customers who had already had their portraits painted. She very kindly promised to approach some of them and, if they agreed, put us in touch.

Back at the hotel, the atmosphere had changed. It was Friday evening and already getting dark. There was a frisson of excitement in the lobby, but I had no idea why. I saw a notice on prominent display requesting that guests respect one another's privacy and refrain from soliciting— with a threat of expulsion for noncompliance. I could hardly believe that this warning was put up with us in mind. Had there been a complaint from "Sugar Kane" about the evening before? I thought it very unlikely, but you never know. My questions were soon answered. Within a few minutes several town cars with darkened windows arrived and disgorged about twenty gigantic men wheeling large suitcases. They didn't check in at the desk but went straight through the lobby to a suite of rooms that looked as though it had been prepared for them. They were soon joined by a more regular-size man who was welcomed by the staff in the most effusive way. He went into the same suite as the others, and that is when I noticed his bright red shoes. I asked our server, Monica from Foster City, what was going on. "They are the 49ers," she explained. "They come here the night before every home game." I knew enough to realize that she was talking about the San Francisco football team, but that was about as far as my knowledge of the 49ers, or indeed the game of American football, extended. "Who is the guy in the red shoes?" I inquired. "That's their chief coach, Mike Singletary," replied Monica in a hushed and reverential tone.

Suddenly, from the direction of the players' suite, a giant of a man, some six feet three inches tall and about 250 pounds, materialized and walked up to the couple at the next table. For such a big man he moved extremely gracefully, with none of the rolling from side to side of the merely overweight. He gave the woman a long and gentle embrace and then sat down for a few minutes of animated conversation before leaving to return to the suite. After the couple had finished eating and left the restaurant I asked Monica who that had been. "That," she replied, "was Ray McDonald." "And the couple?" "They're his parents. They come to every home game," replied Monica. "All the way from Florida." I was amazed at the devotion, not to mention the expense, of a six-thousand-mile round trip every other week. I was also curious, given that Ray and his parents were African American, how they had come to have a Scottish surname.

The following day I settled down in front of the big screen in the restaurant to watch the game while Ulla went for a walk along the promenade by the lagoon. The 49ers were playing the St. Louis Rams. Monica was on duty again and, in between serving her customers, came over to explain what was going on, which was a good thing because until then I could not follow the game at all, with players running in all directions, nowhere near the ball. With Monica's instruction it began to make some sort of sense, the quarterbacks hurling the ball forward, the four attempts allowed to move the ball forward ten yards, and the defense taking out their opposite numbers. With the helmets and the body armor it was impossible to recognize Ray McDonald by his appearance alone, but then I spotted him by his name on the number 91 jersey. No longer the gentle giant of yesterday, playing on defense he charged and blocked with a ferocity that was terrifying. It was a good game for him. He scored a touchdown, and the 49ers won. As well as admiring his play, I could not help wondering whether he also carried the same Macdonald Y chromosome that I first encountered on the Isle of Skye. I felt a tinge of frustration that I had not had the opportunity to find out.

Over the weekend, true to her word, Joanna Mountain had contacted a small selection of friends who had already seen their chromosome portraits, so Ulla and I set about meeting as many as possible. On Monday we were heading down 101 again toward Palo Alto and the hallowed halls of Stanford University. One reason for 23andMe to be located in Mountain View was its proximity to Stanford and the genetics powerhouse it had become under the leadership of one of the true Olympians of genetics, Luca Cavalli-Sforza. It was he who had carried the torch for population genetics from the time of the blood-groupers in the 1950s right through to the modern DNA era. Although I had not always agreed with his conclusions, there was no doubting his pioneering contribution both in research and in writing some of the most influential textbooks. Two of these were my introduction to the field of population genetics and have pride of place on my bookshelf. Practically everyone in the field had been through Cavalli-Sforza's laboratory at one time or another, including Joanna Mountain and my appointment that day, Dr. Roy King.

We drove through the neat main street of Palo Alto and under the arch that joins it to the Stanford campus. This was not a crowded campus of the sort that are all too familiar in England, with buildings tacked on one by one as dictated by the ebb and flow of funding but without any sense of architectural continuity. This was a Tuscan town transplanted from Italy and then nourished in the warmth and prosperity of California. Piazzas flanked by round-arched colonnades, with palm trees shading students reading and talking on the grass. No wonder Luca Cavalli-Sforza, who was born in Genoa in northern Italy, felt so much at home at Stanford.

Ulla and I made our way to the Psychiatry and Behavioral Sciences Building for our meeting with Dr. King, a psychiatry professor there. If I had been a patient suffering from depression, King's specialty, the mere encounter would have taken me halfway to a cure. He was fast-talking, his sentences peppered with bouts of infectious laughter, as he led us to his office on the first floor. King is an African American with short grizzled hair and beard and laughing eyes behind steel-rimmed glasses. Once inside his office he explained that he was from Nashville, Tennessee, where his mother had been teaching biostatistics at Meharry Medical College. Meharry is one of several institutes for higher education founded by religious groups from the North after the Civil War to encourage the enrollment of former black slaves. It still has a largely African American faculty and student body, and is one of the major educators of black physicians in the United States. There she met King's father who had recently returned from fighting in World War II and was enrolled as a medical student. King's grandfather, his mother's father, had been a math professor at Fisk University, also in Nashville, and another of the institutions founded during the Reconstruction era to cater for African Americans. His father, King's great-grandfather, had been a poet, one of the very first to write in the vernacular, and his father, King's great-great-grandfather, had written a diary during the Reconstruction period that was now lodged in the Library of Congress. So on King's mother's side there was a direct line back to the South and the difficult years that followed the abolition of slavery.

In complete contrast, his father had been born in New York and been adopted by the King family. Nothing was known about his father's family until, much later, King and his mother traveled to New York and traced his father's birth certificate. The certificate showed that his mother was an African American with the surname Coleman. Unusually for a couple who were not married, the certificate also named his grandfather. To King's complete surprise his grandfather's name was Maurice Ginsberg, almost certainly a surname with an Ashkenazi Jewish origin. This was the first sign of an ancestry about which King knew nothing. Growing up in Pennsylvania during the late 1960s, when there was still plenty of discrimination, King recalled that he had many Jewish friends. He played golf with some of them at the few courses that allowed blacks and Jews to play. So, before he had his chromosome portrait painted, he knew that at least a quarter of his ancestry was European through his grandfather Maurice Ginsburg. But when he pulled the portrait onto his computer screen, the full complexity of his ancestry was there in green, blue, and orange. Other than demonstration portraits that MacPherson had showed me, this was the first time I had been able to look in close detail at the chromosome portrait of an African American.

My first impression was that it was very colorful, and a lot more interesting than my own almost completely monochrome equivalent. There was about as much blue as green, but with a good sprinkling of orange as well. In fact there was slightly more blue, which indicated that Roy had more European ancestors than he did African. This was considerably more than the 25 percent coming directly from his Ashkenazi Jewish grandfather and, when Roy first saw it, had taken him completely by surprise. I asked how it got there, and he acknowledged that one reason was that, in the late nineteenth century, one of his black ancestors on his mother's side was raped by a drunken white man when she was thirteen or fourteen and became pregnant as a result. This was the first time I had come across this explanation for the European component of African American DNA, but it was not to be the last.

I hesitated to ask King a more difficult question, but then I thought, He's a psychiatrist. He can take it. I wanted to know whether his chromosome portrait, particularly the higher-than-expected European ancestry it had revealed, had influenced how he thought about himself. Sure enough he was not fazed at all, and had already thought about this. He explained that he had grown up at a time when "embracing your African roots" was becoming publicly acknowledged, a time when African American youths did not want to learn anything about European history or culture. He just did not think it had anything to do with him. He only really woke up to his European heritage when he found out that his grandfather was Jewish. His Y chromosome is typical of Ashkenazi Jews, and, like Bennett Greenspan's customers from Family Tree DNA, King has since become interested in Judaism and Jewish culture.

Though he trained as a physician, King had always been fascinated by archaeology and was able to indulge this interest when he took a temporary teaching position at Stanford's outpost in Florence, Italy, for a year. His wife, an art historian, was especially interested in the female figurines of the Neolithic period that are to be found all over Italy and Sardinia. Oddly enough it was this that later drew his attention to genetics, when he read a paper in *Science* about the geographical spread of Y chromosomes in Europe and thought he could see a congruence with one particular haplogroup, J, and the figurines. When he read the long list of authors he realized that some of them worked less than a mile away in Cavalli-Sforza's laboratory on the same Stanford campus. He was at the point in his long career in psychiatry when he wanted to add something else, and genetics became that other dimension. Before long he was working with Peter Underhill, the Stanford scientist who has done so much to discover the genetic markers along the Y chromosome, and publishing on the genetics of the eastern Mediterranean. Only when he tested his own Y chromosome did he realize that it had come from that part of the world through his Jewish grandfather. Was this a coincidence, I wanted to ask, or was it his DNA that had somehow drawn King to feel an affinity for that region? Now we were entering the realm of the Jungian collective unconscious, but I thought once again, He's a psychiatrist. He can take it. "Mmm, it's possible, I suppose," came the noncommittal reply.

While King might have been feeling the homing instinct of his Y chromosome, he certainly agreed that his mitochondrial DNA was pulling him toward West Africa. Like many African Americans, he is a member of the Lingaire clan (L3D), which is found predominantly close to the Atlantic coast of Africa around Senegal. Even before he knew that, he had discovered an affinity with African art, but just not any African art, only art from that region. While he was an avid collector of the colorful abstract paintings from far West Africa, and with examples on the walls of his office, he had no interest at all in pieces from Central or South-west Africa. Was this just chance or an unconscious affinity passed on through his DNA?

There was time for one final coincidence before it was time for us to go. King had used publicly accessible DNA databases to discover an exact mitochondrial match with a man from Togo, which is farther around the Gulf of Guinea from Senegal. He had an unusual surname, Agboto, and when Roy Googled the name he found only one match for it in the entire United States. It was a professor from Togo who was teaching biostatis-tics at Meharry Medical School, just as his mother had done fifty years previously. "Spooky" said King and, with that, our time was up.

The next day, leaving our sanctuary of the airport hotel, Ulla and I joined the 101 once again, but this time heading north toward San Fran-cisco. Our destination was San Francisco General Hospital, where we were to meet with another of Joanna Mountain's nominees, Esteban Bur-chard. Like Stanford, the original style of the hospital was Italianate, with several handsome five-story buildings of multicolored brickwork topped by blind arches. More Venice than Tuscany. But that was where the simi-larity ended. There were no lazy palm-fringed piazzas here. We were in a poor neighborhood and since it was a public hospital, the patient base was largely made up of Americans without health insurance. As we pushed our way through the crowded lobby, the patients queuing for attention were mainly black or Latino. San Francisco General had been the first to treat AIDS patients in the early 1980s. This was well before the disease was understood, and it was the first hospital to witness the epidemic of rare skin tumors, Kaposi's sarcoma, and the unusual lung infections that were the first signs of the immunodeficiency that marked the insidious advance of the retrovirus that caused it. As Burchard explained as he showed us the very ward that took in the first patients, being in a public hospital, the physicians give their time for no financial reward, support-ing themselves instead by research grants through UCSF, the University of California, San Francisco, where he also teaches.

Esteban is a pulmonary physician and an expert on asthma. Mountain had recommended that we meet because in his research Burchard had been using the chromosome paintings to try to understand the very different frequencies of asthma in blacks and Latinos. In fact, his interest in the way gene pools blended went much further back, and he had been a principal author on several important scientific papers on the impact of race and ethnicity in clinical practice that I had read. He had experienced more than his fair share of adverse reaction to his claim that there were genetic differences between races, something that professional geneticists had tried to play down for fear of being condemned as eugenicists. I have said, and written, in the past that my research with mitochondrial DNA showed that race has no genetic basis, but now I think that was an oversimplification. Researching *DNA USA* has given me the opportunity to consider far more closely what this statement means, as I will explain later.

Of all my meetings this one was going to be the most directly related to my old life in medical genetics, and there was a pleasurable familiarity when Esteban began to tell me about his work on asthma. For a start he knew Bill Cookson, a colleague from Oxford, the irrepressible Australian scientist and author of *The Gene Hunters*, who had been searching for "the asthma gene" for as long as I can remember. Both Burchard and Cookson had begun their hunt in the days when, inspired by the triumphs of single-gene disorders, scientists thought there was bound to be only one asthma gene, only to find that their holy grail was a mirage, a goal always just over the horizon. The frustrating truth is that, like so many of the common diseases with a genetic component, there are many genes involved in asthma, with no single one of them having an overwhelming effect. There are many ways to look for asthma genes, and Burchard chose to approach the task by trying to understand the reasons why the incidence varied so much between different populations in America. Asthma is far more frequent in Puerto Ricans than in Mexicans, and reasoning that this difference has something to do with their gene pools, he began to look in detail at the genetic structure of both populations. That is how he came to work with Joanna Mountain and to admire her chromosome portraits, although he had already done a great deal with the AIMs we have already mentioned.

It wasn't long before we were sitting at his computer to have a look at his own chromosome portrait. Like Roy King's, it was a colorful mosaic of the three principal colors—green, orange and blue. But the overall color scheme was very different. Instead of King's mixture of African and European with a bit of Native American thrown in, Burchard's portrait was predominantly Native American and European, with a touch of African. Although Esteban had been born and brought up in San Francisco, both of his parents are Mexican, which explains the orange segments in his chromosome portrait. Like Native North Americans, Mexicans are also descended from the first people who crossed into America from Asia. Burchard had already known that his DNA was a mixture of all three from his AIMs work, although the precise proportions were slightly different from those revealed by his chromosome portrait. Like me, he really liked the visual impact of the portraits and, also like me, saw how they dissolved the anxieties associated with the brutal arithmetic of the AIMs evaluations. AIMs were fine for the kind of large-scale studies that Burchard had carried out in Puerto Rico and Mexico where he was after averages, but far less appropriate for individuals.

I asked what the effect of seeing his own chromosome portrait had been. He was slightly surprised by the amount of European DNA he was carrying, about 40 percent, and assumed, with no obvious disappointment, that his distinctly European Y chromosome had originally come from Spain. This is a common finding among Mexicans, where mitochondrial DNA usually reveals a descent from one of the Native American clans, while among Mexican men more than half carry a European Y chromosome. This is yet another example of the disproportionate genetic input of European men, this time stemming from the Spanish colonisations beginning in the sixteenth century. What meant most to Burchard was the 7 percent of his DNA that was African. Having done a lot of his asthma work work among African American communities in

Puerto Rico and Boston, he now felt far more closely connected to them and to their struggles against the discrimination they faced. He reflected that had he lived a few generations before, during the days of the Jim Crow "one-drop" rule, where even the tiniest fraction of African ancestry meant you were relegated to a life outside the Caucasian mainstream, his life would have been utterly different as a result of the smudges of green in his genetic portrait. As it was, he now felt much more at home with his African American asthma patients. Their response to his newly revealed African ancestry was one of two alternatives: "No wonder we liked you" or "No, you're not."

There were a few more people that I wanted to see in the Bay Area, but they would not be available for another week, which meant a bit of rescheduling. The original plan, such as it was, had been to make the return journey to the East Coast by train from Los Angeles to New Orleans, then to Washington after a few days in Atlanta, Georgia, which I had wanted to visit to get a flavor of the South. If we were going to stick to that itinerary, we could not wait another week in San Francisco. Ulla and I went down to the hotel lobby once more to think about it. As well as hosting the normal tourist guests, the hotel was also a thriving conference venue with labeled delegates arriving from all over the United States. One day it was a big travel company, the next it was the sales force from Victoria's Secret. That day it was the turn of a well-known pharmaceutical company. We were seated at a table on a sofa with another one at right angles and, presently, two young women, both African American, sat down next to us. They were laughing and joking and, quite soon, Ulla slipped herself effortlessly into the conversation. I sat there smiling mildly in a professorial sort of way until the usual "Where are you guys from?" routine was over. I didn't usually listen to the opening moves, but this time I could not help overhearing that the pair was from Atlanta. By the time I was properly introduced, they were eager to know more about the research—and I was eager to oblige—especially given where they were from. So that is how we encountered "Mildred Pierce" and "Ned Land." (By now I was running low on available and gender-appropriate

Hollywood pseudonyms for my volunteers, and women were having to adopt masculine monikers.)

"Mildred" and "Ned," it transpired, were both colleagues and friends. "Mildred" was a production manager, while "Ned" was a corporate lawyer in the legal department. The conference was over, so they were relaxing before their flight back to Atlanta the following morning. Both had been born and brought up in the South, and both, as far as they were aware, had only African American ancestors. They knew that their ancestors had worked on the plantations and had no idea where in Africa they might have come from. They were glad to have a DNA test, so we withdrew to one of the booths, and moments later the samples were taken and soon on their way to the lab. Meeting "Mildred" and "Ned" had been a delight, but it also meant that we would not miss out completely if our new schedule left no time for a visit to Georgia. If we could not get to Atlanta, then in a way Atlanta had already come to us. This also gave us enough time to meet everyone Mountain had suggested and fit in a side trip to Houston to visit Bennett Greenspan, the founder of Family Tree DNA.

That evening, the no-soliciting notice went up again, and I asked my football tutor, Monica, why that was. Even though only a week had passed, the 49ers had another home game. Sure enough Mr. and Mrs. McDonald arrived from Florida and as before their son Ray came to say hello to them. After he left, Ulla homed in (in the nicest possible way), and before long we had joined Ray senior and his wife, Labrina, at their table. They were absolutely charming, and we talked about their son's football career from high school to college and then as a professional. And, yes, they did come to watch every single home game and, yes, it was a long way, but well worth it. We soon began to discuss what it was that we were doing in San Francisco, and they both became very interested. On the origin of their McDonald surname, they had no idea how they had come to have it but they would quite like to know whether they were related to the chiefs of the eponymous Scottish clan. I explained the procedure, in this case a simple cheek swab from Ray senior and a Y-chromosome test. I went to get a kit from our room while Ulla kept on talking. Back at the table, the coffees were arriving. At that point the coach, Mike Singletary, came in, still with his red shoes, and greeted them briefly before moving quickly off in the direction of the players' suite. I can't say for sure if the coach's presence had been the immediate cause, but I could see that Ray senior was having second thoughts when he said that he would need to talk to his son. I knew the chance was lost. But never mind, we had all enjoyed a good conversation, one we would not have had were it not for our research. The next day the 49ers went down to a heavy defeat.

Among the many friends and relatives whose portraits Mountain had painted, genetically speaking, one stood out from the rest. Not in her attributes as a friend, but rather in her chromosomes. Whereas most came out colored in the expected hues, Caucasian blue or African green, or a mixture of all three, like King and Burchard, "Ilsa Lund"'s chromosomes came as a genuine surprise. "Ilsa"'s ancestry, as far as she knew at the time, was monotonously British—"bog standard" was the term she used. In that respect her European background mirrored many of Mountain's customers who had purchased the DNA tests principally for the health information they contained. When Mountain first saw "Ilsa"'s chromosome portrait she was so excited that she called her immediately. That had been only a few days before, so when "Ilsa" met us at what had become our usual table in the hotel lobby, the news was still fresh. She had brought her chromosome portrait with her and immediately put it on the table. I could see at once what Mountain had found so surprising. Among the blue chromosomes were long tracts of African green. Not just a small fleck but really long stretches covering four chromosome arms. In numerical terms, almost 8 percent of her DNA had a recent African ancestry.

"Ilsa" told us that her family had been in the Carolinas from the early days. She had not yet tracked down the immigrant ancestors, but all the signs were that most or all of them had arrived in the seventeenth or eighteenth century. The only inkling of something other than a completely European background was that one of her great-uncles had claimed, without much evidence to back it up, that there might have been a Delaware Indian somewhere in there. Given her brown hair, very slightly olive skin, and brown eyes, "Ilsa"'s coloration would not have stood out from that of the crowd back home in England. Nevertheless, in the past she had been taken for a Cuban, a Puerto Rican, a Mexican, an Israeli, and even an Eskimo. Not so her sister, who according to "Ilsa" was a blue-eyed blond.

Since she had just learned of her African ancestry, I was interested in her immediate reaction to it, before it had really sunk in. She was certainly surprised, but only because from what limited knowledge she had of her ancestors, none of them had been African. If anything, she said, she might have expected some American Indian ancestry because of her great-uncle's theories. She would have welcomed that outcome because she felt, like many other Americans I met, that some Indian blood would be a good thing to have, a connection to the original Americans. She certainly had no strong adverse reaction to finding that she had an African American ancestor, or any very positive one either. Judging by the long lengths of uninterrupted green chromosome segments, there was probably only one and, on the doubling-dilution-per-generation rule, he or she would have joined the genealogy about three or four generations ago. A great- or great-great-grandparent, in other words. After her initial surprise, "Ilsa"'s next puzzle was to identify this unexpected ancestor. She had told her cousin Greg about this, and he, being an avid genealogist, was on the case. Her own theory was that her African ancestor might have joined the family tree around the time of Reconstruction, but she promised to update me on any developments as she and Greg dug into their past. A year later she let me know that further tests on her relatives had confirmed that her African American ancestor was on her father's side, but that was as far as they had gotten. I had the distinct impression that "Ilsa"'s relatives were not so eager as she was to find out more.

When we met in San Francisco I had asked "Ilsa" whether the news of her ancestry had any effect on her attitude toward African Americans, maybe those she knew through work. "Ilsa" works in the IT department of one of Silicon Valley's large biotech companies, and she said that most of her colleagues were Indian (from India, not America) or Chinese. Only a minority were European, and there were only two African Americans she could think of. One irrelevant detail that caught my attention was that, despite this being California, the only Latinos in the company were on the janitorial staff. So she didn't know enough African Americans to feel any immediate empathy with her newfound cousins. She had not booked a trip to West Africa yet, and would definitely not be checking the African American box at the next census. She hadn't told her elderly mother yet, but her blue-eyed blond sister's reaction was telling: "It's nothing to do with me," as if it were an African ancestry that "Ilsa" had inherited but she had not. Obviously "Ilsa"'s African ancestor would also be her sister's, and barring an extreme statistical freak, they would have roughly the same amount of African DNA.

Our next appointment in the Bay Area was also one of the most intriguing. "Will Kane" met us in the lobby, and we settled down at one of the tables away from the hubbub and ordered a late lunch of soup, half an eggplant sandwich, and a bottle of New Zealand Sauvignon Blanc, one of my favorites. As we tackled our food, she (we were getting very short of predetermined pseudonyms by then) began to tell us about herself. The first thing I noticed about "Will" was her aura of calm and detachment. It was noticeably different from that of "Ilsa Lund," or of "Mildred" and "Ned" from Atlanta. She told Ulla and me that she had been born and raised on the Navajo reservation near the Four Corners region, where the state lines of Arizona, Utah, New Mexico, and Colorado all come together. Her mother was a Navajo, although her distant ancestor had been abducted from the nearby Hopi reservation. Her grandmother spoke only Navajo. Her father was an anthropologist who had come to the reservation, met and married "Will"'s mother, and stayed. Her home had been near the small town of Kayenta, Arizona, not far from Monument Valley. It is a very rural region, very dry, and with many people raising sheep for a living. Both her parents worked as language teachers, and "Will" had attended high school in Kirtland, a two-hour bus ride from Kayenta, just over the border in New Mexico.

She had always enjoyed science in high school, and when she enrolled in college at the University of Arizona in Tucson, she soon found herself working on a genetics project on albinism, the pigmentation disorder in which melanin production is faulty, resulting in very pale skin, hair, and eyes. Albinism is present among the Navajo and other Southwest Indian tribes, and "Will"'s supervisor set her the task of finding out whether the mutation, in a cell-membrane protein, that caused it was the same in the Navajo as had been found in other tribes. To do this "Will" had collected DNA samples from her own tribe, though she had been careful to double-blind the results so that she never knew the identity of the donors. To cut a long story short, she did find a new mutation among the Navajo that was subtlely different from those in neighboring American Indian tribes. She thought that the likely reason why the mutation was confined to the Navajo was because their numbers had been severely reduced during the long conflicts with European settlers, starting with the Spaniards, and then the European Americans. Although there were now more than a quarter of a million Navajo, the tribe had been reduced to about five thousand after they had been forcibly interned at the tiny Bosque Redondo Reservation near Fort Sumner, New Mexico, in the mid-1860s. The reservation was completely unsuitable, and many Navajo died before they were allowed home in 1868. In "Will"'s opinion this was the event that squeezed the population down to a small bottleneck through which the particular mutant version of the albinism gene had passed and then flourished as the population increased.

"Will" had done this work around the time of the ill-fated Human Genome Diversity Project, of which more later, that had resulted in the tribe issuing a moratorium on genetic testing on the reservation. Being herself a Navajo, "Will" had been allowed to collect the samples she needed for her albinism research, but only outside the boundaries of the reservation. After she graduated from the University of Arizona, "Will" came to Stanford as a graduate student and started work on the pigmentation genes of the mouse, a popular model organism for genetic research that can then be extended to understanding more about humans. The project had worked out very well, because the week before we met, "Will" had successfully defended her Ph.D. dissertation at Stanford. One of her advisers had been Joanna Mountain, which is how we came to be introduced.

Outside, a pelican was flying across the lagoon between the hotel and the airport runway, periodically diving headlong into the water in search of supper. I could see its crop extend as it gulped in a mouthful of seawater and then expelled it, trapping and swallowing any edible creatures that it contained. Inside we ordered some more sandwiches and another bottle of wine. Our conversation turned back to her experience of the Navajo attitude to genetic testing. "Will" was about to begin postdoctoral research into the ethical issues involved. She wanted to overcome the resistance that had developed because she could see the health benefits that might follow from Navajo participation. But everything that had happened, the Havasupai case and the Human Genome Diversity Project, had made the Navajo very wary of genetics, and of science in general. Just as Serle Chapman had told me about the Cheyenne, there was understandable indignation over cooperating in genetics projects only to be told that your long-held tribal beliefs about your origins were wrong.

I spoke again to "Will" after I sent her a copy of her chromosome portrait. She was making some progress in her conversations about genetics with the tribe, and was making a formal study of the impact of the Havasupai court settlement, but it had been hard work. She had been perfectly happy to take part in my research for the book, and we had taken a DNA sample for her chromosome portrait. With her European father and Navajo mother, it would be quite easy to distinguish between the chromosomes that she had inherited from each of her parents. But what fascinated me was that, while she was very interested in what the DNA might reveal about the ancestry of her European father, she really did not want to know about her mother's contribution. It was as if the enthusiasm for genetics shown by my European volunteers and the resistance to it that I had witnessed among Native Americans were living side by side in the same individual. Her paternal chromosome wanted to know while her maternal chromosome did not.

My last meeting in San Francisco was to catch up with someone I had met in Oxford a few months before setting out for the United States when he had been among the guests at a dinner at my college in Oxford. The conversation at our table soon turned to what work we were each doing and, when it was my turn, I explained that I was researching a genetics book on America. At that point one of the guests surprised me by revealing that he was a Cherokee Indian. I looked as closely at him as the low light, and manners, allowed. He certainly seemed to have some of the features I had been led to expect, including high cheekbones, dark hair and eyes, a proud and distinctive nose, and slightly darker than usual skin tone. Although he was based in Oxford, he explained that he was going back to the United States for the next two semesters at Berkeley, and that he was leaving the very next day.

I realized long ago that Oxford college dinners had an unwritten rule. Though conversations might indeed be fascinating and engrossing, even intimate, no promises made on these occasions need ever be kept. It is the university equivalent of "we must do lunch sometime." However, on this occasion we did keep in touch and arranged to meet in San Francisco. Which is how Ulla and I came to be driving over the Bay Bridge to Oakland and up the freeway toward the Claremont Hotel on the hills behind Berkeley. This location was much closer to the UC campus than our airport hotel, and I welcomed any excuse to return to the terrace restaurant with its fabulous views back across the bay and to the Golden Gate Bridge in the background. "Lucas Jackson," his new pseudonym, was waiting for us and had already ordered one of the Claremont's renowned ruby grapefruit cocktails. The sun was hovering over the bay, and the heat of the day was slowly fading. As another two ruby cocktails arrived, we sat back to enjoy the view. The lazy thud of tennis balls coming up from the already floodlit courts below combined with the first effects of the ruby specials to lend a very relaxing tone to the reunion.

Although we were meeting in San Francisco, "Lucas Jackson" had been brought up farther south, in Los Angeles. Even so, here I was, on the farthest west coast of America, meeting someone whose Cherokee ancestors had traveled, over many generations, three thousand miles from the other side of the continent. As "Lucas" started to talk about his ancestors and what they meant to him, I looked more closely at his face in daylight than I had been able to in the low artificial light of a college dining hall. He is tall and slim, with dark eyes and noticeably dark eyelashes. His hair is dark brown too, but closely cut. His skin pigmentation is a little darker than mine, but he in no way resembles the images of American Indians of either my imagination or of photographs I had seen. By now a plate of spicy calamari had arrived, and my recorder captured "Lucas"'s words between mouthfuls:

"The Cherokee were originally from Georgia and South Carolina. They were one of the five so-called Civilized Tribes, the others being the Creek, Choctaw, Seminole, and Chickasaw, I think. They got the name because they were already living a settled agricultural life when they encountered the earliest European settlers. They lived in long rectangular houses, had their own governments and legal systems, wore cotton clothes, and grew corn and squashes on irrigated plots.

"At first relationships with the Europeans were pretty good. There was trading and intermarriage. Nevertheless the Cherokee kept themselves apart from other Indian tribes, which is why we look different, so I am told.

"The Cherokee had their own language, of course, but they were unique in that they became the only American Indian tribe to have a writing system. This was all down to one man, Sequoya. Like many Indians, Sequoya, who was a silversmith, was impressed by European writing. He called it 'talking leaves.' Single-handed, he produced a system of letters and symbols to represent words around 1820. At first he had a lot of difficulty persuading fellow Cherokees to adopt his new system. He traveled to Cherokee settlements in Oklahoma and Arkansas to explain it. It is still used there today. Sequoya's dream was to reunite the fractured Cherokee Nation through language but, after a crusade through Arizona and the Southwest, he died in Mexico, trying to link up with dispersed Cherokee who had moved there. And yes, the giant redwood was named after him, or so I believe.

"Back in the Carolinas relations between Cherokee and the Europeans began to deteriorate as the British Americans, in particular, wanted to expand their colonies westward further into Cherokee territory. This was in the early 1800s. The encroachment was sporadic to begin with. There was no deliberate policy of mass displacement, at least not yet. However, as land first rented by Europeans was then purchased, my ancestors were gradually pushed westwards. I don't know whether there was any organized armed resistance, but I suppose that when persuasion failed there would have been raids, then counterraids. That would have provided the excuse for the forced evictions. You can see it happening. At first my ancestors were displaced into the mountains, the Appalachians, where the land was much poorer than in South Carolina.

"After the Louisiana Purchase at the start of the nineteenth century, vast territories were taken over by the United States. Thomas Jefferson, the president, thought there was plenty of room in these new territories for all the Indians who were living east of the Mississippi. In the opinion of the Congress, the so-called Civilized Tribes had made such good progress toward integration, that they seriously considered establishing an Indian state in the new territories and admitting it to the Union. But the southern states, Georgia in particular, demanded more Indian land in order to expand the plantations. They threatened to secede from the Union if they couldn't get it. So Congress passed the Indian Removal Act around 1830. The main driving force behind the act was the man who signed it into law, President Andrew Jackson, whose face is on the twenty-dollar bill.

"The act called for the removal of all Indians living east of the Mississippi to the new U.S. territories on the other side of the river.

Although many Cherokee set off westward immediately, my ancestors, along with most of the tribe, lacked the means to do so. They were eventually evicted and forced to march on foot. This was the notorious 'Trail of Tears.' Four thousand Cherokee died on the journey, the soldiers who drove them along the way refusing to allow them to stop to tend to the sick or even to bury the dead.

"Eventually the survivors, my ancestors among them, were settled in Oklahoma, where some of my relatives still live. Of course, this wasn't empty land and had its own population of Plains Indians, mainly Cheyenne and Arapaho. I can remember my great-grandmother, who died when I was seventeen, saying how she regarded the Plains Indians as 'wild' and nothing like her. They were hunters and lived in tents, while the Cherokee had always been farmers. She said she had much more in common with the whites than these wild Indians of the plains. I remember my great-grandfather, who died when I was about five, because of his false teeth, which he used to snap loudly to scare me and my brother. He denied being an Indian at all, even though he spoke the Creek language, played stickball games, and went to stomping dances, which are a basically a pow-wow.

"In the 1930s, during the Great Depression, farming in Oklahoma was hit hard by the dustbowl conditions, when all the topsoil was blown away. My grandparents, like many 'Okies,' had to move west again, to the Central Valley of California. This was the same journey that John Steinbeck wrote about in *The Grapes of Wrath*. You probably remember Henry Fonda in the film even if you haven't read the book. My grandparents, my father's mom and dad, both settled in the very same town, a town called Weedpatch, believe it or not, near Bakersfield. They got work picking onions, strawberries, and carrots but my grandfather eventually landed a job as a state trapper. The work was catching coyotes and other vermin. It wasn't glamorous, but unlike the picking, it did pay regular wages and a he got a small pension when he retired.

"After my father met and married my mother, whose ancestors were

European for as far back as I know, they moved to Sauglas, which is just north of Los Angeles. That's where I was born and grew up. It is only an hour's drive from my grandparents near Bakersfield, so I went to stay with them a lot. My parents moved back there when I left for college.

"When I was a child I always thought that my mother's family were much more fun than my dad's. Her father was a big-game hunter, and the house was full of taxidermy, mounted heads of deer and so on. But as I have gotten older, I have begun to appreciate the qualities of my father's Cherokee side of the family. These are, I would say, a great sense of personal and social responsibility, not relying on others to help, but at the same time helping others where they could. My grandmother, who died at eighty-eight and lived alone until the last month of her life, volunteered at the local social center. There was something unshakable about her, a quiet fortitude that surrounded her. The same was true of my grandfather. He was something of a trickster, a practical joker. He told me about when he went to a neighbor's watermelon patch and hollowed out the insides of the fruit, then hid and waited for them to discover what he'd done.

"He told me about 'noodling' for catfish back in Oklahoma when he was young. When catfish lay eggs, they collect them into a sort of mud nest that is usually hidden under a bank. One fish stays behind to guard the nest, and they will defend it. Noodling involves diving into the river and literally fighting with the catfish, trying to get a hand into the gills, or in through the mouth and then through the gills, and heaving them onto the bank. Catfish are strong, so it is quite dangerous, but a big fish, which could be five or six feet long, would feed the family for quite a while. During the Depression this was what kept them going."

By now, with the calamari well on the way to digestion, another round of cocktails had arrived and the sun had slid into the ocean behind the Golden Gate. The sky took on the bright orange glow that I have only ever seen in California, and the first stars appeared overhead. We talked of many things until the air had cooled enough for the servers to light the patio heaters. The lights had come on all over San Francisco and, illuminated by their headlight candles, the great unending river of cars flowed smoothly across the Bay Bridge toward us. The silhouette of the Golden Gate slowly melted into the night sky, leaving only the red aircraft warning lights to mark its presence. Eventually even the gas heaters could not keep the cold away, and all three of us made our ways home.

16

The Call of the Canyon

Our original schedule was looking rather ragged by now, but the fortunate encounter with the two African American women from Atlanta meant that we had a few days to spare before we needed to be in Washington, D.C. After my intriguing conversation with "Will Kane" I realized that the time could be put to no better use than in paying a visit to the Navajo reservation where she had been brought up. Not that I had any intention of even trying to obtain any DNA samples, I just wanted to see what it was like. We booked a flight to Las Vegas, picked up a car, and headed off on the long drive to Flagstaff, Arizona, past the Hoover Dam and then through scrubby desert peppered by faded settlements with evocative names like Grasshopper Junction and the distinctly chemical Chloride.

It was dark by the time we arrived in Flagstaff, and our first accommodation, on the outskirts, was a bad mistake. I had been away from home for two months now and long exposure to hotel air-conditioning had given me a chesty cough, which was beginning to feel and sound like a recurrence of the asthma I had years before. This got a lot worse in the

Monument Valley.

desiccating atmosphere of this particular building, created by a croaking air-conditioning unit and exacerbated by the extreme dryness of the outside air and the elevation. Flagstaff is almost seven thousand feet above sea level. Besides, since the hotel was a set of suites, there was no food and we had to make do with a very basic Thai meal three blocks away. Not what you want after a long drive. Early the next morning I went in search of somewhere else to stay, preferably without air-conditioning, and headed into downtown Flagstaff. This was a very different place from the outskirts, with two-story brick houses laid out in small blocks. On one corner stood an old hotel, the Weatherford, and it had a cancellation. Within an hour we were on the upstairs veranda drinking beers and listening to old Beatles tracks coming from the loudspeakers on the walls above our heads. My dry cough retreated a little further with every familiar track.

As any American road buff knows, Route 66 passes through Flagstaff. From there the original road takes the steep ascent to Oatman and on to the promised land of California, ending on the Pacific coast at Santa Monica, 2,448 miles from its starting point in Chicago. There are plenty of notices in Flagstaff alerting you to the town's location on this iconic highway, but if you miss them, the throaty roar of Harley-Davidsons cruising up and down soon lets you know. We had a great view of these shiny chrome machines from the veranda as they slowly growled their way along the narrow streets before roaring off to rejoin Route 66 near the railroad tracks. Very occasionally another make of motorcycle, a Triumph or a Yamaha, puts in an appearance only to be "seen off" by the Harleys and scuttle down the side street, out of town.

The Weatherford Hotel opened its doors in 1900 and since then has had its fair share of famous guests, including the much-traveled President Theodore Roosevelt and his friend Charles "Buffalo" Jones, the first game warden of Yellowstone and a hunter-turned-conservationist who tried to interbreed buffalo with cattle, which didn't work. However, the project did take Jones on a fund-raising lecture tour back east during 1907. In the audience in New York was a dentist from Ohio whose name was to become the one most associated with the Weatherford. His name was Zane Grey, and he was so enthralled by Jones's tales of the outdoors, the mountain lions, and the adventures that he gave up dentistry and began to write for a living. He became the most prolific novelist of the American West, writing more than ninety books with worldwide sales of forty million. His first books were his best, but although his later works were rather repetitious, his readers still loved them. In later life, by then a millionaire, Grey traveled the world indulging his first love, which was fishing.

Zane Grey was also instrumental in publicizing New Zealand as an unspoiled mecca for fly fishing, which is how I first heard his name, at Turangi on the Tongariro River near Lake Taupo on North Island. That was where Richard caught his first wild trout. Zane Grey often stayed at the Weatherford on his research trips to the West, so I did not mind at all following in his footsteps. To complete the experience I walked around the block to Starlight Books on North Leroux Street and bought a copy of *Call of the Canyon*. I spent the rest of the day on the veranda in the company of the rugged Glenn Kilbourne, living rough in the Arizona wilderness after returning from World War I, and his reluctant fiancée, Carley. Reluctant, that is, to leave New York and the soft life of cocktail parties and shopping that made up the daily routine on the Upper East Side for wealthy young ladies in the early 1920s. Having recently finished my namesake Plum Sykes's debut novel *Bergdorf Blondes*, I was struck that nothing much seems to have changed in the intervening ninety years. In the end, of course, Carley relents, and true love transports her to the West where she and Glenn live happily ever after on the range.

The fortuitous move to the Weatherford was amplified by the evening's session in the bar by a Celtic band, which was just as passionate and lively as anything on the Isle of Skye. In the mid-nineteenth century Flagstaff was growing just as Scotland was exporting Highlanders in the Clearances, so there has been a strong Scottish community here ever since.

While it was charming to be reminded of home, it was not getting us any nearer the Navajo or Hopi reservations. Our first step was to stop at the excellent visitors' center near the train tracks. We looked through the brochures, but while there was plenty on the nearby Grand Canyon there was nothing much about the reservations. Then a young man with a broad smiling face came up to ask if we need any help. He told us he was half Hopi himself and pointed out the best way to get to the reservation. Ulla wanted to enroll him as a volunteer straight away, but I was reluctant to put him on the spot, so Ulla and I went back to the hotel for lunch. Over a Reuben sandwich of corned beef, sauerkraut, and melted swiss cheese—Zane Grey's favorite apparently—we wondered what we should do about our new Hopi friend. I was still reluctant about overstepping an invisible line, especially after talking to "Will Kane" about the Navajo moratorium on genetic testing, and reading about the even greater reluctance of the Hopi to reveal themselves. But Ulla's enthusiasm was, as usual, persuasive. In the end I agreed that she could go back to the visitors' center and see if our friend would like to join us for a chat at the Weatherford. I don't know how she did it, but an hour later we were all sitting in the hotel lounge and ordering beers.

Although I certainly did not expect it, our friend did want us to do a DNA test. This meant he had to have a pseudonym, and the next film character out of the bag was "Roger Thornhill." The newly anointed "Roger" told us that both of his maternal grandparents were Hopi. His grandfather had been born before any proper records of births or deaths were kept, so he didn't know when he was born or how old he was. This left him free to choose his own birthday, and he settled on the Fourth of July because, as "Roger" told us with a smile, this way he was guaranteed fireworks on his birthday. His grandparents had been brought up in different villages and belonged to different clans. His grandmother was a member of the Spider Clan, while his grandfather belonged to the Sun Clan. A close relationship between members of different clans was frowned on, so they eloped and left the reservation to live and work together at the Grand Canyon. From there they moved first to Phoenix, then to Flagstaff in the 1960s. They had saved enough money to buy a plot of land and build a house, where "Roger"'s mother had been born. His grandfather is still alive and, at around eighty-three, now spends most of his time back on the Hopi reservation where, over the years, he also built a house. "Roger" told us that poverty is still a big factor on the reservation, with many people living without electricity or running water. Generating income is still a major problem, and though some still farm the land, those with cars look for work in the closest towns, like Winslow, Gallup, and Flagstaff, which also operate a shuttle to and from the reservation.

I could tell that "Roger" was used to helping people: Our conversation was punctuated by sudden thoughts about what he thought might interest us. The Hopi have lived in the same villages for longer than any other Indian tribe, with Old Oraibi on Third Mesa being the longest continually occupied village in the whole of North America. As I had picked up from my visit to the Cheyenne, there was a deep division within the tribe between traditionalists on the one hand, who are reluctant even to share their history with the outside world and just want to be left alone, and modernists on the other, who are eager to embrace the life they see outside the reservation and often decide to leave. This is not just a benign ideological debate. It is a real clash between colliding ideals, which, as "Roger" told us, has recently become violent with gas stations on the reservation—symbols of the new—blown up and destroyed.

The other long-running battle has been with the Navajo, whose reservation entirely encircles the Hopi land. The two tribes have quite different ancestral origins, reflected in their completely different languages. While the ancestors of the Hopi, so it is believed, had moved up from Central America at least two thousand years ago, the Navajo had migrated south from northwestern Canada much more recently. "Roger" could tell the difference by their appearance, the Navajo being taller and slimmer compared to the shorter, more compact Hopi. In "Roger"'s opinion the Navajo were more progressive than the very conservative Hopi, and that attitude had helped them to expand their territory and surround the Hopi lands.

The key moment was in 1901, when a Navajo delegation arrived in Washington, D.C. to lobby President Theodore Roosevelt to grant them

more land so as to reduce the frequent clashes with the growing numbers of white settlers in and around Flagstaff. The Navajo hired a cler-
gyman, the Reverend William Johnston, to come with them, not so
much for himself but because his nine-year-old son, Philip, was fluent
in Navajo and acted as their interpreter. Philip was later to be instrumental in proposing the use of the Navajo language as the basis for the
secret code used by the U.S. Marines in the Pacific during World War
II. But back in 1901, what a scene that must have been with a child acting as the vital link between the Navajo nation and the president of the
United States. (In much the same way, the young John Quincy Adams,
later the sixth president, had acted as French interpreter for the American mission to the court of Catherine the Great in St. Petersburg in 1781,
when he was only fourteen.) With Philip Johnston's help, the Navajo
intervention worked, because Roosevelt signed the Leuppe Extension
Treaty, which sealed the expansion of the tribal territory at the expense
of the now-encircled Hopi. I had read enough about Navajo history to
know that this was a tremendous simplification, and that the president's
apparent generosity was in stark contrast to the terrible treatment the
Navajo had received at the hands of earlier administrations, especially
the mass deportation to Fort Sumner, New Mexico, in 1864, that was
only reversed after a public outcry.

To return to "Roger"'s own story, his paternal grandfather had been the
son of a Mohawk father and an Ojibwa mother. He had faced the familiar
prospect of discrimination, but being light skinned, he managed effectively
to conceal his Indian ancestry, adopted a familiar Irish name, and lived
as a white man. Rather like Zane Grey he was drawn to Arizona, and his
love of Westerns led him to enroll at the University of Arizona in Tucson,
and there he met and married "Roger"'s grandmother, a redhead European American with a Dutch-German background, and then they moved
to upstate New York, where "Roger"'s father was born. So, from a genetic
point of view, "Roger"'s ancestry was one-quarter European American and
three-quarters Native American. "Roger" told us that he planned to go to
New York one day and find out more about his Dutch-German forebears.

Unlike "Will Kane," "Roger" had not been brought up on the reservation but in Parker on the Colorado River, just south of Havasu, close to the border with California. (It was also incidentally, the final resting place of a structure I knew as a child—the same London Bridge I used to cross to meet my father after he had finished work in the City of London, when we would go for a meal and a movie. The old bridge of 1831 had been moved across the Atlantic, numbered stone by numbered stone, after its demolition in 1967.) It was while the family was in Parker that "Roger"'s mother was asked if she would like to join CRIT, the acronym for the Colorado River Indian Tribes, a federation of Mohave, Cheme-huevi, Navajo, and Hopi. She declined even though the federation owns and operates the BlueWater Resort & Casino in Parker, which adds millions of dollars to the federation's coffers and allows generous educational grants to its members, which would have included the young "Roger." As it was, "Roger"'s family moved to Prescott, Arizona, from where he enrolled in Northern Arizona University at Flagstaff, where he has stayed ever since.

"Roger" was an intriguing mix of the traditional and the modern. He felt himself to be Hopi at heart, but was also completely at ease with the modern world. There was none of the quiet, almost demure reserve I had sensed in "Will Kane." "Roger" was open and optimistic. He had dated Navajo girls while at high school but had found them too bound by tradition for his taste. In college he found himself hanging out with the white students, which is how he met, fell in love with, and then married his wife, Emily, a red-haired girl with her origins not among the arid mesas of the Hopi and the Navajo, but in the emerald meadows of far-off Ireland. At the same time "Roger" feels deeply attached to his ancestral roots in Hopi land even though he never lived there. But he is also very aware of the titanic ideological struggles going on in his ancestral homeland.

Finally "Roger" left us with instructions of how to get to the Hopi mesas, which is where we headed the next day. I knew there was no prospect of any DNA collection on the reservation, and I had no intention of even trying to do so. Indeed, on each occasion that I had entered an

Indian reservation I made quite sure the DNA sampling kits were left behind. This was primarily out of respect, but it also removed any residue of temptation I might have felt when I got there. I was, after all, a collector of genes, and I was well aware of the devil that comes over all collectors, be it of fossils, archaeological artifacts, or butterflies, banishing all sense of danger, moderation, or propriety when surrounded by the objects of their desire.

We headed east along Highway 40, which is hereabouts superimposed on the old Route 66, toward New Mexico. The tracks from *Easy Rider* were replaced by a collection of cowboy songs featuring, among others, "Rawhide," "Riders in the Sky," and "The Streets of Laredo." I need only put them on now, and I am back on the long, dusty road. On each side the stands of ponderosa pine that surround Flagstaff thinned out then disappeared, leaving a flat semidesert of dried grass and yellow shrub stretching to the horizon under a cobalt blue sky. Tumbleweeds really did blow across the road. Other than the highway and the occasional roadside signs urging voters to back McCain and Palin, left over from the previous year's presidential elections, there was nothing else. Gradually the grass cover thinned and was replaced by reddish brown grit dotted with stones and isolated tufts of grass. There were no cactuses or Joshua trees to alleviate the dry monotony. Eventually hills appeared on our left horizon. These were the Hopi mesas, our destination.

After the desolate town of Winslow, flanked by boarded-up cafés, we headed north up Highway 87. As the country got even drier, the hills on the horizon began to separate and take shape as we approached. Now there were ramshackle farmsteads and small herds of black cattle on either side of the road. But what they found to eat I could not imagine. There was nothing there, even less than when Richard and I had crossed the Great Basin of Nevada on the train, some weeks back. The road ahead stretched to the ends of the earth, or so it seemed. The hills became banded cliffs of gray and delicate pink; then, a little farther on, dark buttes like the cones of small volcanoes erupted out of the desert.

They were not really volcanoes but the last remnants of a great plateau of sedimentary rock that had been eaten into by the wind and left standing. In time they also would also be ground into dust and disappear. Still the road stretched out ahead, a gray ribbon laid across the desert. The buttes were left behind, and the road headed straight for a low cliff on the horizon. This was Second Mesa, home of the Hopi for at least the last two thousand years. As we got nearer we could make out white dwellings on the steep sides and along the top of the cream-colored limestone cliff. Once the Hopi had also lived on the arid plains, but they had moved all their villages up onto the mesas following the Pueblo revolt of 1680 as a defensive tactic in anticipation of Spanish reprisals.

Warned by "Roger" not to take any photographs, Ulla put the camera away when we reached the base of the cliff, and we followed the road as it wound upward past the old village of Shongopori and onto the plateau at the top of the mesa. From here you could see how the three Hopi mesas stretched down like fingers from a much bigger and darker formation to the north—Black Mesa, one of the causes of friction on the reservation, as we shall see. We turned right along a narrow ridge that led to two more villages, Shipaulovi and Mishongnovi. A thousand feet below us the butte-studded plain stretched out into the distance, while on the ridge itself there were small patches of corn and squash. What a contrast to the endless dense green stands of corn and soybeans the Zephyr had passed through on its way across the Midwest. The houses were made of mudbricks and adobe around dirt squares. A man carried a pet ground squirrel, and three teenage boys were playing soccer.

We had come to the Hopi mesas not on a spiritual pilgimage but just to see what they were like—to experience the atmosphere, what people we saw, and of course the all-important landscape that is such an integral part of the Hopi spirit. Without these visits, my wonderfully revealing conversations with "Will Kane" and "Roger Thornhill" would lack any narrative context, and their genetic results would mean less as a result.

A few miles down the road we were lucky to find the modest but atmospheric Hopi Cultural Center. I felt reluctant to ask too many questions, as I would have under normal circumstances, and walked slowly around the displays of old photographs, ceramics, and the strange kachina dolls. These are the representations of spirit beings that visit the mesas during the spring and early summer before returning to their spiritual homes among the San Francisco peaks above Flagstaff. Without soliciting inquiries, a woman busied herself in the small office. It was only when I offered to buy a small book about the Hopi that she revealed herself as Anna Silas, both the book's author and the founder of the museum, which she had opened twenty years before.[1] Then Ulla started a conversation, and although I did ask a few questions I felt slightly as though I was intruding on a secret world where I had no business to be. Having now read Anna's book, I realize that it was not really the center's intention to be secretive, but neither was it there as an entertainment. Rather it was providing a glimpse into a world where not everything was meant to be completely revealed to other than the Hopi themselves. We met, or rather saw, more Hopi in the small adjacent café, where the servers said all the right "American" things but were palpably gentle and restrained. They reminded me of the Polynesians I had met in the South Pacific, and like them, they were also heavier than they should have been, a harbinger of the diabetes that afflicts all Pueblo tribes in the Southwest.

As we headed back down the sides of the mesa toward our own temporary home below the San Francisco peaks, it struck me that we had glimpsed a way of life not dissimilar to that of all our ancestors. A way of life that had been lost a very long time ago but one that nonetheless had shaped and molded, through millennia of evolution, the DNA we still carry. While the life of the Hopi on Second Mesa might seem very foreign and strange, and while very few of us would willingly return to it, even were it possible to do so, it resonated with a deep ancestral memory that we all share. Although it was easy to retain that thought as we headed back under the darkening sky, it got a lot harder when we were finally reimmersed in the chaos of modern life.

The following day we headed north toward Kayenta and the heart of the Navajo reservation. Once again the ponderosa pines thinned and disappeared as the highway headed for the pinks and yellows of the Painted Desert. We passed through the sprawling trading post of Cameron and on toward Tuba City, the most populous on the Navajo reservation. The road was dotted with shacks advertising shawls and jewelry, none of them inviting. Tuba City itself was elegant in comparison, with the accoutrements of any small American town—that is to say a large gas station and a McDonald's, just beyond which a modern school was decanting its neatly turned out pupils into the ubiquitous yellow school buses. Right along the route to Kayenta, fifty miles to the northeast, the buses spread out in all directions along unseen tracks in the desert, their routes traced by clouds of thrown dust. Their destinations were the scattered clusters of dwellings, mostly trailers or small bungalows, that were home to the Navajo. I had read that the Navajo find it difficult to live in close proximity to one another, and this scene confirmed it as the dust plumes moved out into the desert for miles in every direction.

The land was certainly dry but not as utterly arid as the Painted Desert. I discovered later that an aquifer ran deep underground between Tuba City and Kayenta, and wells kept the residents well supplied with water. Or at least they did once. As Ulla and I neared Kayenta a new railway line curved in from the north and ran along the road. Where can this be going? we wondered. The answer lay a few miles ahead when we saw a giant tower. As we got closer we could see a conveyor belt running high above over the road toward the tower, under which the rail tracks fanned out into a classification yard. This was, as I was later to discover, the exit route for coal from the Black Mesa we had seen in the distance from the Hopi villages the day before. The whole hill was made of coal, and it was being systematically reduced. The presence of the mine was and still is highly controversial—one more symbol of the internal struggles within the Navajo nation. Leased to one of America's largest energy companies, the same one that Richard and I had seen ripping into the ground on our way to Bear Lodge, the coal and other minerals, including uranium, located nearby are a valuable source of income for the Navajo, but it comes at at price. The negotiated terms of the lease were particularly advantageous to the company, and questions about its propriety still hover. Until a few years ago, before it was shipped off by rail, the coal was pulverized, then mixed with groundwater, and the slurry sent through a pipeline more than 250 miles to the generating station at Laughlin, Nevada. If that seems a crazy thing to do in a desert, then it is. Even six years after the slurry operations were closed down, the people of Kayenta still can't get enough water from the aquifer.

Kayenta itself is less developed than Tuba City and had one sure sign of poverty, groups of dull yellow stray dogs, ribs protruding from their emaciated bodies. Why, I asked myself, are stray dogs always yellow? They looked exactly the same as the ones I had seen scavenging in Polynesia. There was no reason to stay long in Kayenta, and we left town heading toward Monument Valley. Buttes and mesas lined the route, glowing orange in the setting sun. It was dark by the time we reached our accommodation, a former trading post converted into a lodge. Monument Valley and its spectacular scenery, hidden from us for the moment by impenetrable darkness, came to the nation's attention only when the owner of the trading post persuaded John Ford, the Hollywood director, to use it as a location for his Westerns. Many of these movies starred John Wayne, and the small hut that the star had used as his home on location was right next to the lodgings. Right next door to that, a small theater was showing his movies every evening.

Once we had settled in, Ulla and I made our way over to see that evening's screening of Wayne's *She Wore a Yellow Ribbon*. Under normal circumstances I would have sat through the film, even though Westerns are not my favorite genre. But as soon as the stereotype "Indians" appeared whooping and hollering, I thought how crassly insensitive it was to screen such a film on a reservation. Ulla, who is always one step ahead of me when it comes to summing up a situation, got up, grabbed my hand, and led me straight outside. Of course, on reflection it was just another sign of the dilemma faced by the Navajo, like the Black Mesa mine. Acting in a John Ford Western such as this, even as a caricature of yourself, was a rare opportunity to earn cash in this utterly beautiful but unforgiving land.

I cannot pretend that our visit to Monument Valley was undertaken entirely in the cause of research. I awoke early and looked out the window. The sun was just above the horizon. Before me a wide plain of the familiar scrubby desert stretched for a few miles into the distance. On the horizon stark mesas and vertical-sided buttes glowed a soft pink in the early sunlight. This was not Monument Valley itself—that lay over a rise a couple of miles away—but even so the scenery seemed very familiar. I knew full well that whole area had once been covered by the deep and ancient sediments of the Colorado Plateau and that the sculpted monuments were all that was left of these thousand-foot accumulations of limestone and sandstone after first water and then wind had gouged and blasted the hardened rock over millions of years. But that isn't what they looked like. Instead of being the eroded cadaver of a once-great plateau, each isolated cliff, each pinnacled mesa seemed like a high castle built up from the ground long, long ago. All that was missing were the princesses imprisoned at the top of every one.

Like all other tourists we took the drive around Monument Valley itself, where the formations were all the more dramatic and all the more familiar from the movies. Afterwards we sat on the terrace of the newly built hotel on the valley rim. By then the sun was going down behind us and the shadows were beginning to creep up the petticoats of fallen rock that surround the base of each rose-red monolith. The hotel was built and staffed by Navajo, and the way our server quietly and courteously brought our order had the same calmness and grace that we had seen in "Will Kane" back in San Francisco. At a nearby table a group of four middle-aged women sat chatting, their floral tops and permed hair slightly ruffled by the breeze coming up from the valley floor. From the pulses of overheard conversation that wafted across to our table, the women were

Mormons from Salt Lake City. A ground squirrel darted out to grab a crumb from beneath their table and sped away back to his burrow under the sagebrush. By now the shadows had climbed almost to the top of the rock in front of us, and its shadow was projected for miles across the flat desert beyond. A dust plume betrayed the car that made it. Far, far away.

These places are made for contemplation, and my thoughts returned to the Navajo and how they had come to be here. As the murmur of conversation floated across from the other tables I became aware that, within the few square feet of the terrace, three completely different myths existed side by side. My own unashamedly rational and scientific version, based on the evidence of genetic links with Siberia and the Pacific; the women from Salt Lake City, who were sure the Navajo and other Native Americans were descended from the Lammanites of Israel; and the Navajo who worked here who believed their ancestors had left the fourth world through a rent in the sky and entered this, their fifth, through the gasping, steaming vents of Yellowstone.

Just as the sun disappeared below the horizon, the desert was suddenly flooded with a strange purple wash. The fairy castles glowed bright terra-cotta for a moment, then were consumed by darkness.

17

A Question of Blood

After a side trip to Houston to see Bennett Greenspan from Family Tree DNA, our next rendezvous was in Washington, D.C., where we were to meet up with Gina Paige, the cofounder, with Rick Kittles, of African Ancestry. With a day to spare until Paige was back in town, like so many visitors to Washington we made our way past the White House and down to the National Mall. On previous occasions I had been to most of the museums which line the wide avenue that leads from the Lincoln Memorial to Capitol Hill, but I had never been inside the most recent addition. In prime position, nearest to the Congress buildings, lies the National Museum of the American Indian, its curvilinear architecture and golden limestone outer facings deliberately recalling the weather-eroded mesas and buttes of the West. Pools and artificial wetlands surround the entrance, and the whole building reinforces the intimate connection with the land that is so important to indigenous Americans. Everything about the museum has been guided and led by Native Americans. Inside, as in nature, there are no sharp

The Buffalo Dancer by George Rivera, National Museum of the American Indian, Washington, D.C.

edges, and the spacious atrium rises all the way to the top of the building. A gentle ramp connects the four floors that are home to permanent exhibitions based on the original collection of Native American artifacts assembled by the avid collector Gustav Hey, while temporary rotating displays feature particular Indian communities. A sure favorite with visitors, and our immediate destination, was the Native Foods Café on the ground floor, which divided into regions each offering its own specialties. Ulla and I headed for the Northwest Coast and its delicious servings of hot-smoked wild salmon.

As well as viewing the magnificent collections I was searching for any reference to genetics within the museum. I wanted to see how the curators were dealing with the new information about the ancestry of American Indians and whether it threw any light on the distrust and hostility toward genetics that had been ignited by, among other things, the Havasupai case. I could find nothing. I asked at the front desk in case I had missed anything, but was met by blank stares. After I returned to England I got in touch with Gabrielle Tayac, one of the NMAI curators, to ask her more about the museum's current thinking about genetics. She confirmed that there is no mention of it within the museum but that it was something they are thinking about very hard. They are in a difficult dilemma.

According to Dr. Tayac, one reason for the antipathy toward genetics is that it is perceived as an affront to traditional knowledge, knowledge of the kind that Serle Chapman had explained to me during our days in the Bighorn Mountains. Genetics is seen as the latest way of invalidating native traditions that had been going on in one way or another for centuries. Native traditional religions were suppressed in the United States until the passage of the American Indian Religious Freedom Act of 1978, which recognized that Indians had been denied the right to freedom of religion enshrined in the First Amendment. The act also acknowledged past infringements by the federal government, such as restricting access to religious sites, and also, as this accompanying statement from President Jimmy Carter concedes:

In many instances, the Federal officials responsible for the enforce-ment of these regulations were unaware of the nature of traditional native religious practices and, consequently, of the degree to which their agencies interfered with such practices. This legislation seeks to remedy this situation.[1]

Against this background, it is easy to appreciate the potential for hostility to genetics among Native Americans. The concession that federal officials were unaware of the effects of what they were doing could just as easily be applied to scientists in the Havasupai case and others like it. Indeed some proposed genetics projects in the past are guilty of the same insensitivity, one that even I could see. None demonstrates this so well as the Human Genetic Diversity Project from the early 1990s.

As the head of steam was building for the Human Genome Project, the massive undertaking to reveal the entire sequence of human DNA, leading population geneticists saw the opportunity of adding a satellite project to study genetic variation around the world. There was nothing particularly new in the concepts behind the project—after all that is what they had all been doing for years. But the genetic tools were new, and the scale would be vast, which appealed to a certain type of megalomania not uncommon among geneticists at the time. What made me doubtful was not the grand scale, although my personal preference has been for less ambitious research projects, but some of the reasoning behind it. In particular I was disconcerted by one stated aim, which was to take samples from isolated populations before they disappeared. I had a startling vision of a small native group standing around their campfire and looking up to see the Land Rover coming over the horizon loaded with dry ice (to freeze the samples), and knowing that their time was up.

Another feature of the Diversity Project was the health benefits it promised, in the sense that greater understanding of human genetic variation might get to the causes of diseases in general and those that were endemic to some groups in particular—like diabetes among American

Indians, for example. This was a realistic ambition at the time, but critics of the project pointed out how often in the past indigenous people had been recruited to medical research projects that held out the prospect of future health benefits that never materialized. There were problems with the commercial patenting of genes, very popular in those days, and there was a danger that others might benefit financially from discoveries made through indigenous people. Some of the wording used by the Diversity Project was certainly clumsy. Describing indigenous people as "isolates of historic interest" was patronizing to say the least. Intentionally setting out to "develop a panel (of DNA markers) for forensic identification of individuals" raised fears that this would be used to develop methods of forensic investigation based on ethnic origins. Against a historic background of suspicion, this was asking for trouble.

The taint of scientific racism that wafted around the Diversity Project still haunts the collective memory, especially, but not exclusively, among Native Americans. Critics of the project were concerned about a direct effect: that if DNA were used to define racial groups, then some governments might use genetic data to discriminate against them. As the Jews had realized some time before, it was a theoretical possibility, even though the correspondence between the two is very loose indeed. Fears were even voiced that biological weapons could be developed that targeted particular groups based on their DNA. For all these reasons, which were never adequately countered by the organizers, the funding dried up and the project never got off the ground. In my view it was on shaky foundations from the start, not so much because it was such a bad idea but because of the intoxicating menace of hubris that surrounded the whole enterprise.

But in one sense the Human Diversity Project lives on. In April 2005 a consortium of the charitable Waitt Foundation, *National Geographic*, and IBM launched the Genographic Project. This was certainly a project on a grand scale, backed by the considerable resources of the organizations behind it. Its aim was to collect DNA from indigenous people all over the world to trace the history of human migration. I had heard about the

project before it launched and eagerly sent away for my copy of the intro-
ductory DVD, which ironically for a worldwide enterprise was in a format
that restricted it to U.S. audiences. The Genographic Project had learned
from the failure of the earlier Diversity Project and sensibly avoided the
most contentious issues. There was to be no medical data collected and
the genetic analysis was restricted to the two genetic lines with which we
are already familiar, mitochondrial and Y-chromosome DNA.

Yet there were undeniable similarities between the two projects,
particularly in the composition of the organizing committee, whose
chairman was the same distinguished Stanford geneticist, Luca Cavalli-
Sforza, who had inspired the original project. The work of the Genographic Project was divided up among several international academic
centers with long experience of population-genetics research, in other
words many of the same people who were hoping to run the Diversity
Project. According to the publicity the results of the Genographic Project, after peer-reviewed publication, were to be be put into the public
domain.

By now aware of the enormous public interest in individual results,
the project also offered a public participation kit by which, for a modest fee, interested members of the public could submit their own DNA
for analysis. The proceeds from the sale of these kits were used to create a Legacy Fund, which aimed to put something back into participat-
ing communities by way of grants designed to increase awareness of the
threat to indigenous cultures.

All this sounds very positive and a far cry from the original Diversity
Project. Judging by the number of fee-paying public participants, well
over three hundred thousand to date, the Legacy Fund will be well pro-
visioned. The Genographic Project is very well resourced and attractively
presented, as you would expect from *National Geographic*, and it should
be forgiven for the rather disingenuous original claims that it was something entirely new. It certainly is from the point of view of public awareness, but from a scientific standpoint it is really only replicating a decade
or more of genetics research that had already been done, and published,

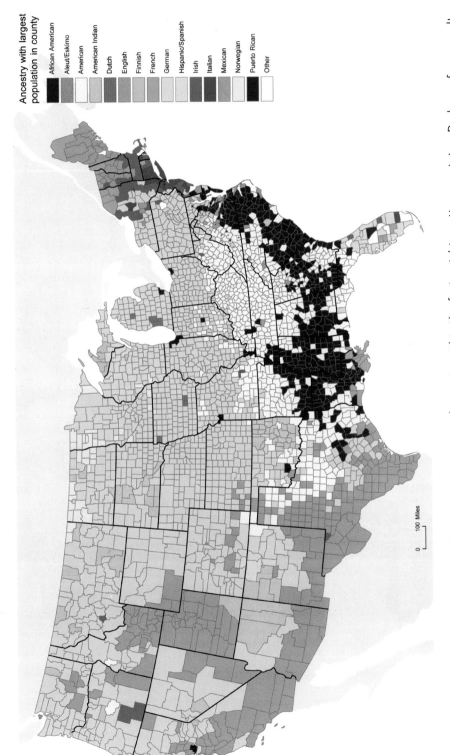

Ancestry with largest population in county

- African American
- Aleut/Eskimo
- American
- American Indian
- Dutch
- English
- Finnish
- French
- German
- Hispanic/Spanish
- Irish
- Italian
- Mexican
- Norwegian
- Puerto Rican
- Other

0   100 Miles

Self-declared ancestries with the highest proportion in each county within the forty-eight contiguous states. Redrawn from results of the 2000 census.

THE *DNA USA* CHROMOSOME PORTRAITS GALLERY

"MARGO CHANNING"*

"ATTICUS FINCH"*

"HARRY LIME"*

BRYAN SYKES

Chromosome portraits showing the ancestral origin (green: African; blue: European; and orange: Asian/Native American) of DNA segments within the twenty-two autosomes of twenty contributors and volunteers. Fuller details are in the text. Asterisks indicate pseudonyms.

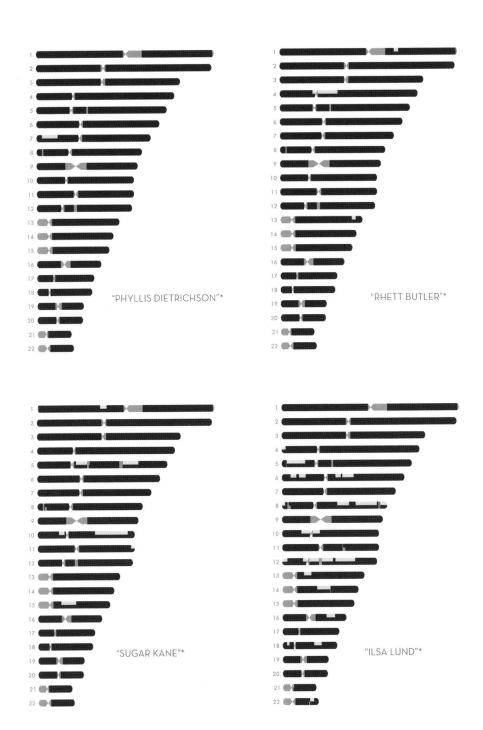

"PHYLLIS DIETRICHSON"*

"RHETT BUTLER"*

"SUGAR KANE"*

"ILSA LUND"*

TOBY COOPER

"MILDRED PIERCE"*

"NED LAND"*

RICK KITTLES

All images © 2012 23andMe, Inc.

"VIRGIL TIBBS"*

"HOLLY GOLIGHTLY"*

MARK THOMPSON

"NORMA DESMOND"*

All images © 2012 23andMe, Inc.

HENRY LOUIS GATES Jr.

"WILL KANE"*

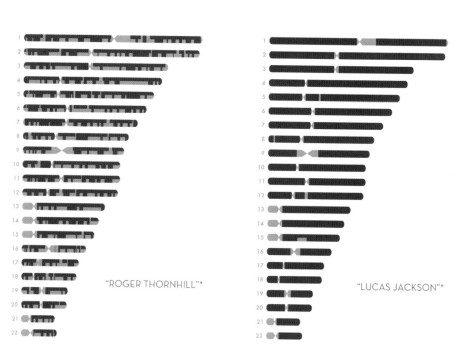

"ROGER THORNHILL"*

"LUCAS JACKSON"*

All images © 2012 23andMe, Inc.

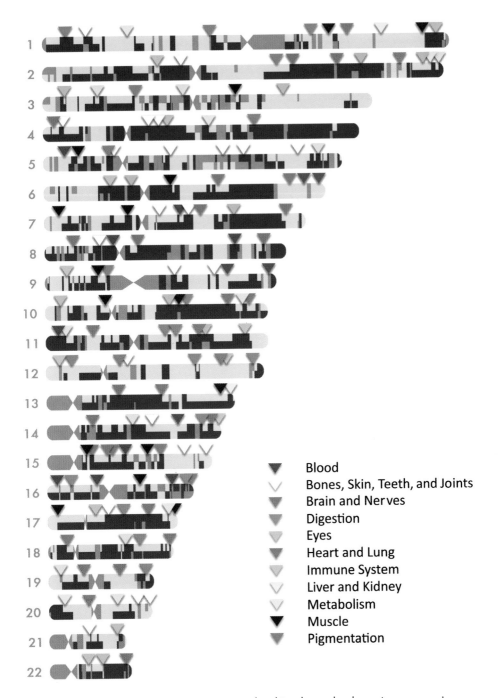

Locations of 146 important genes involved in eleven body systems superimposed on the chromosome portrait of Mark Thompson and showing their ancestral origins (green: African; blue: European; and orange: Asian/Native American). Where locations have two colors, Mark has inherited genes with different origins from his parents. Details of each gene are in the appendix.

much of it by the contributing laboratories. Though I look forward to reading the conclusions and the peer-reviewed publications when they appear, I don't expect them to alter the pretty good picture we already have about how our ancestors moved around the world. But the presentation will, I'm sure, be absolutely stunning.

For all its good points, the Genographic Project still has its critics, which is the feature I want to examine in greater detail as they get to the heart of the difficult situation in which Native Americans find themselves. We have already touched on the historical reasons for the suspicion of the motives of genetic researchers—the clumsiness of some projects and the way they are perceived as attempting to invalidate Native American traditions. To explore this further I contacted Kimberly TallBear from the University of California at Berkeley. Dr. TallBear has been a vociferous critic of the Genographic Project, genetic testing of Native Americans, and—I think it is fair to say—genetic testing in general. It was her remarks that DNA testing is intrinsically racist that I wanted to explore.

She kindly sent me her publications. I had already read one opinion piece in *Science*, published in 2007, with Dr. TallBear among the contributors.[2] In it she and her coauthors warned against the hidden dangers of commercial genetic ancestry testing and concluded by recommending its close regulation. Unsurprisingly I disagreed, and I did submit a response to *Science* that, also unsurprisingly, was not published. But that is hardly relevant here. In another article Dr. TallBear examines the claim that I and others have made in the past—that modern genetic analysis refutes any biological basis to racism.[3] In this context I have been asked to undertake comparisons of Catholics and Protestants in Ireland, Jews and Arabs in Palestine, and Hutu and Tutsi in Rwanda, all with the intention of reducing the tensions between them by demonstrating that they are closely related genetically. Naïve intentions, perhaps, and I did not carry out any of them, but I imagined that the results would come to that conclusion—and in the case of Palestine, studied by others, they did. But as we are all aware, the appreciation of a shared ancestry has not led to any easing of the conflict between Israel and the Palestinians. So although this use of genetics may have been ineffective in terms of its influence on conflict, and may eventually change attitudes ever so slightly for the better, I never thought of it as harmful. It came as an unpleasant surprise to read that it was seen by some to be just that.

In the past the methodology of population genetics did, inadvertently, create false boundaries between groups of people. It originated from a time when the only genetic data that were available came in the form of frequencies—typically of blood groups or similar genetically controlled features. To make any sense of the data, there was no alternative to dividing the subjects into "populations." After all, individuals don't have blood-group frequencies. The "populations" were often defined by such factors as geographic location or language and, lo and behold, there were frequency differences between them. I never liked this kind of analysis, because I thought it created artificial biological divisions on the basis of genetic data. Happily, it wouldn't have been any use for designing biological weapons, or even for placing an individual in his or her right "population."

One of the refreshing changes introduced by the individually precise genetic data now available, largely through mitochondrial and Y-chromosome DNA, is that this type of group analysis is no longer necessary. And yet, I am afraid, it persists in many scientific publications. So, insofar as any segmenting of humans into groups on genetic grounds could be interpreted as a precursor to supporting a genetic basis for racism, then I think it might. Once again it is entirely unintentional, but as we have already seen, that does not make it harmless.

Whether DNA can ever be used to define a person, or his or her affiliations, has become, and will continue to be, a contentious issue for Native Americans because it touches on the very current issue of tribal enrollment. Rather like the forced adoption of surnames in England in the thirteenth century, the initial reason for trying to sort out who belonged to which tribe was purely administrative. It followed the Allotment Act of 1887. This piece of legislation is usually regarded by historians as an attempt to undermine the traditional way of life of the American Indians and their system of communal landholdings. Briefly, tribal lands held in common were to be broken up into 160-acre plots held by individuals, and any unassigned plots could then be sold by the government. Not only was this a misguided attempt to turn Plains Indians into yeoman farmers, it also played into the hands of speculators after cheap land, who then bought up vast areas from the Indians at the going rate of fifteen dollars cash for the whole plot. The Allotment Act had a devastating effect on Indian life. Close families were given plots a long way apart. The land, intended for agriculture, was often barren and useless. Land sharks, like James W. Denver (after whom the city was named), moved in to snap up bargains, and by 1920 two-thirds of the land previously under Native American title—that is, nearly 100 million acres or the area of North and South Dakota combined—had been transferred to European Americans.

Whatever the original intentions or the ultimate effect of the Allotment Act, its administration did call for a register of tribal membership. This was the origin of the Dawes Rolls, named after Massachusetts senator Henry L. Dawes, who sponsored the Allotment Act of 1887, and now widely known as the Dawes Act. The provisions of the act were first applied to the "Five Civilized Tribes," the Cherokee, Choctaw, Creek, Chickasaw, and Seminole, who had been resettled in Oklahoma after being forced from their lands in the Carolinas and Florida during the 1830s. The Dawes Commission that organized the rolls was soon flooded with applications, even though the act itself was extremely unpopular among tribal members. Many refused to apply, some were forced to do so, some left for the rugged Cookson Hills in the east rather than enroll. Nonetheless the Dawes Commission received a quarter of a million applications between 1898 and 1907, when it closed. It accepted a hundred thousand of these on the basis of genealogical relationships, by birth or marriage. Even so, there were many whites with no Indian ancestry who tried to pass themselves off as Indians in order to get an allotment of land. Despite this and other irregularities, the rolls are still relevant today, as it is often a requirement for new applicants for tribal membership to prove descent from someone on the Dawes Rolls.

Although, of course, genetic tests were not available at the time, there is an inference of some biological ingredient with the insistence of genealogical descent as a criterion for acceptance onto the rolls. This was made more explicit in the troublesome concept of "blood quantum" to determine eligibility for tribal membership that followed the Indian Reorganization Act of 1934. This was a much more benign piece of legislation that was championed by one of the more enlightened commissioners of the Bureau of Indian Affairs, John Collier, serving from 1933 to 1945 under President Franklin D. Roosevelt. The act reversed the Dawes privatization of commonly held land and by so doing returned two million acres to Indian custody. It also began the process of transferring legal powers to tribal councils—which are still in force today. Though far from perfect, it was a step in the right direction. Although succeeding administrations have tried periodically to dismantle tribal structures in order to assimilate American Indians as fully integrated U.S. citizens, these efforts have routinely failed.

Today membership of a recognized tribe is a highly sought-after privilege. American Indian nations now make their own rules for tribal membership, and almost all contain a reference to "blood quantum." This refers to the proportion of an applicant's ancestry that is considered to be tribal. So, if one of your parents is a "full-blooded" member of the tribe and the other is not, your blood quantum is one-half. The minimum requirements vary among tribes, from one-half in the case of the White Mountain Apache of Arizona, to one-sixteenth for the Mashantucket Pequot of Connecticut and one thirty-second for the Cherokee Nation. Just as the original Allotment Act encouraged fraudulent applicants after land, so has tribal membership attracted a deluge of new applicants eager to enjoy the financial and other benefits it confers. There are the reparation payments for illegally acquired land, educational grants, subsidized health care, income from the lease of mining rights, and, most prominent of all, revenues from gaming. The most famous example of the last, Foxwoods Resort Casino, is owned and operated by the Mashantucket Pequot and has the largest turnover of any gambling resort in the world, dwarfing both Atlantic City and Las Vegas, a remarkable achievement for a tribe that numbered only fifteen members in 1900 and whose landholdings had shrunk to less than an acre.

Since a landmark legal case of 1905, tribal authorities have been able to define their own membership criteria, and most do use a version of the blood-quantum estimation. They are not obliged to do this, but it really only reiterates the requirement for a genealogical link to other tribal members, past or present. Of course, the use of the term "blood" is emotive, but other nonbiological epithets would do just as well, even if lacking the same immediate psychological impact. Nonetheless, though the complete autonomy of tribal-membership decisions substantially dilutes the charge of a racist element imposed from outside, it does leave open the possibility that genetics could be used to back up an application for tribal membership in the absence of sufficient genealogical proof. However, some critics of blood-quantum estimates see them as dictated by outside forces, even if operated by the tribes themselves. "All tribal enrollment efforts obliged the descendants of Native people to think about where they fit in a white-dominated, racialized world."[4]

DNA has certainly already been used to establish an applicant's paternity, if only to reinforce a genealogical claim. But that has nothing to say about defining a tribe through DNA. I know from my own experience that DNA-testing companies are regularly approached by aspiring applicants for tribal membership for genetic "proof" with which to back up their claims. This is technically not possible, as tribes cannot be defined by DNA, nor should they be. Not that this fact will stop unscrupulous commercial outfits from pretending that they can.

Attempts have been made, though, by groups of people to use genetics to apply for federal recognition. The Western Mohegan tribe in Vermont used rather old-fashioned methods of tissue-type frequency in an attempt to demonstrate a link to federally recognized tribes, and found one in Wisconsin. This failed to pass the state legislature, but not until after the organizers had been accused by some other tribes of "enabling" genocide.[5] More recently, in 2005, the Seaconke Wampanoag from Rhode Island approached the Genographic Project looking for genetic evidence to fill in gaps in the knowledge of their ancestry where written records are in short supply and oral histories have been largely lost.[6] It was a brave step by the tribal chair and genealogist, Michael Markley, who told me that much thought, prayer, and counseling by the elders was required to allay the anxieties of those tribal citizens who found it hard to accept genetic testing. But, as he added, "bravery is in our genes." The Seaconke are one of several groups within the Wampanoag people, only two of which, the Aquinnah from Martha's Vineyard, and the Mashpee from Cape Cod, are recognized by the Department of the Interior and thereby entitled to benefit from federal assistance programs. However, as Michael Markley emphasized to me, the Seaconke have never asked for federal recognition, and this was not the motivation of their participation in the Genographic Project.

The name Wampanoag means "People of the First Light" in their Algonquian dialect. They once inhabited much of the eastern seaboard of New England and were the first Native Americans to be encountered by the *Mayflower* settlers in 1620. By then their numbers had already been severely reduced by an epidemic, probably smallpox, said to have been introduced by French fur trappers. They famously helped the English settlers through their first winter and participated in the first Thanksgiving celebration in the fall of 1621. However, in 1675 they had tried unsuccessfully to eliminate the Puritan settlement at the Plimouth Plantation. After this defeat and the retributions that followed, tribal numbers dwindled to less than four hundred.

The results of the Genographic Project study were published in July 2010 and make very interesting reading.[7] On the matrilineal side, all of the mDNA lineages are of either European or African origin, while the patrilineal Y chromosomes show a range of Native American, European, and African lineages plus one surprise from New Guinea. However, genealogical reconstruction showed that the single Native American Y chromosome was most likely introduced into the tribe by a Cherokee incomer several generations back. This mixture of origins for the Y chromosomes is a reflection of the extensive admixture experienced by the Seaconke since the seventeenth century, and in this part of America with such a long history of contact, probably unremarkable in its composition. In contrast, the complete absence of Native American mDNA among the Seaconke Wampanoag came as a great surprise to me, given the usual direction of intermarriage between African and European American incomers and Native American women. I should point out that an analysis of the autosomes by chromosome painting might well reveal very substantial Native American ancestry among the Seaconke Wampanoag that was missed by mitochondrial DNA, but somehow I doubt it.

In one way this result demolishes the genetic definition of race, as applied to the Seaconke Wampanoag at any rate. It shows the multitude of individual ancestral origins, which, taken either individually or together, would never be able correctly to assign anybody to the tribe with any conviction. As such it is a vivid demonstration of the impossibility of doing the same in *any* Native American tribe. At the same time I thought it must be disappointing for the Seaconke since, at first sight, it hardly supports the notion of direct biological descent from the original tribal members. And yet it is a true reflection of the genetic makeup of the tribe. However, Michael Markley, one of the paper's authors, reassured me that, even though I might have been surprised, he considered the results from the Genographic Project were consistent in every way with the Seaconke's developed ancestry and oral history.

When Richard and I had visited Rick Kittles in Chicago, he had told me about his experience with another group whose appeal to genetics to back up a claim to the entitlements of tribal membership had also produced an ambiguous result. Back in 2004, before the time of chromosome painting, Kittles had been asked to help members of the Freedmen Association of Oklahoma fight their expulsion from the Seminole Nation three years earlier. As my Cherokee volunteer, "Lucas Jackson," had recently explained to me, the Florida Seminole were one of the five tribes forcibly relocated west of the Mississippi in the 1830s.

Like the other "Civilized Tribes," the Seminole had kept black slaves and after Emancipation in 1863, the freed slaves became full tribal members. They were treated as equals by the tribe with no boundaries to intermarriage, with the freed slaves—the Freedmen—often marrying the children of their former masters. What seemed to be a model of racial inclusion continued for more than a century, as the editor of the local newspaper the *Seminole Producer* remarked in a revealing account in *Wired* magazine: "You've got Indians marrying whites, Indians marrying blacks. It was never a problem until they got some money."[7]

The money that put an end to decades of harmony was a combination of gaming receipts accrued since the passing of the Indian Gaming Regulatory Act of 1988, and American government reparations for historic land seizures. For the Seminole the reparations alone amounted to $56 million, according to the *Wired* account. Once the money began to pour into the tribal coffers, applications for tribal membership mushroomed. For example, membership of the Cherokee Nation of Oklahoma, which also saw a similar increase in its financial fortunes, rose from fifty thousand in 1980 to more than a quarter of a million today. In 1983 the Cherokee introduced a requirement for tribal members to carry a "Certificate of Degree of Indian Blood" that was open to anyone who could establish a genealogical link to the Dawes Rolls. As we have already seen, the rolls were not directly concerned with any literal concept of a blood relationship, but the very title of the certificate infers a link based on biology. When the Seminole expelled all two thousand black members of the tribe in 2000, it was not long before the Freedmen realized that directly establishing at least a degree of authentic Native American ancestry through a DNA test should help them challenge the expulsions where other legal means had failed.

When Kittles heard about this from a friend at the University of

Oklahoma, he paid a visit to a conference of the "Descendants of the Freedmen of Oklahoma," the organization dedicated to ending "discrimination against people of mixed Indian African descent," and offered free DNA ancestry tests to delegates. This was before the refinement of the chromosome portraits was available, and the tests were based on the less sensitive AIMs technology Kittles had developed with Mark Shriver of the University of Pennsylvania. Even so, the AIMs tests would give a figure, however crude, for the degree of Native American ancestry in the Freedmen's DNA, which could be compared to other groups of people. Kittles tested a total of ninety-five Freedmen descendants and presented his results to an expectant audience at their 2005 annual conference. He had found that the proportion of African ancestry among the Freedmens' descendants ranged from the lowest at only 4 percent to the highest at 76 percent. Likewise there was a large range in the proportion of European ancestry among the group, ranging from zero to 62 percent. When it came to the most keenly anticipated result, the component of their genomes that was assigned to Native American ancestry, the figures fell between zero and 30 percent, with an average of 6 percent. This was, as Kittles announced to the hushed audience, roughly the same as the average African American from Baltimore or New York City.

Even though this was a lower average component of Native American ancestry than many descendants had hoped for, several declared there and then that they would use the results to press for tribal membership. But the prospects for success do not look bright, judging by the reaction of the Seminole leader, Jerry Haney, who was behind the expulsion of black tribal members in 2000. When asked by the *Wired* reporter if he would reconsider his stance based on new DNA evidence his reply was brutally frank: "They can claim all the Indian they want," he said, "but they cannot become a member of the Seminole Nation by blood. They're down there [on the roll] as Freedmen. They're separate." Nor does litigation offer much hope to disenfranchised Freedmen descendants as complete tribal autonomy in matters of membership has repeatedly been upheld in the courts, up to and including the U.S. Supreme Court.

The original source of the problem traces back to the way the all-important Dawes Rolls were compiled back in 1906 by a delegation of mainly white clerks sent out from Washington. It was their task to vet applications for tribal membership and entitlement to the landholdings that were being distributed. Applicants were directed to one of the tents that had been put up, one of which dealt with the Freedmen. The enrollment system was bound to be arbitrary, and in the absence of documentation or the ability to speak an Indian language, several applicants found themselves on the Freedmen's Roll based on a clerical assessment of their physical appearance. The fact that siblings were sometimes assigned to different rolls only emphasizes the intrinsic fallibility of the system. In those cases, one sibling would be issued a blood-quantum certificate while the other on the Freedmen's Roll would be denied one, even though their ancestry was exactly the same.

What is missing in all of this is a parallel measure of Native American ancestry among the enrolled tribal members, which has not been done. Kittles is not alone in suspecting that such a test would reveal that they, too, had a very mixed genetic background, and that many of today's enrolled members of the Five Nations would have lower proportions of Native American ancestry than some descendants of the Freedmen.

18

Portraits of America

Gina Paige was able to squeeze us into her busy schedule the following day, just as she was setting off again to address a conference in Charleston, Virginia. To make things easier, we met in the refurbished splendor of the concourse at Washington's Union Station, the most magnificent railroad terminus of our trip so far. The gilded ribs of the barrel-vaulted roof reflected the soft white floodlights and classical statues graced alcoves around the second-floor balcony. Gina met Ulla and me in one of the many concourse restaurants and, after ordering lunch, we got down to business. It was Halloween and the station was full of young men and women in elaborate costumes, which livened up the atmosphere but made it hard to hold a conversation without shouting.

Unlike Rick Kittles, the scientific brains behind African Ancestry, Paige had a business background with experience in several commercial areas, the last working for a bakery products company. She was frustrated that, in any of the major sectors, there were no truly new products. Innovations were merely minor modifications of existing products—a

Detail from bronze of Phillis Wheatley by Meredith Bergman. Boston's Women's Memorial, Commonwealth Avenue, Boston, Massachusetts.

line extension or a new flavor whose marketing depended on exaggerating their basically trivial advantages. So when she was introduced to Kittles, she immediately recognized that genetic genealogy was a genuinely new market and that excited her. Combining their complementary skills, Kittles and Paige set up African Ancestry. Like so many African Americans, Paige felt passionately about her own community and was very aware of the sense of disconnection many black people felt. She could also see the potential to assuage that feeling by reinforcing their African origins through DNA. From the start, African Ancestry had a strong sense of mission.

Just as with Bennett Greenspan in Houston, all of us had shared some common experiences from customers, but Paige's mainly African American customers were more anxious about what happened to their DNA than Greenspan's largely white clientele. In her opinion, this had a lot to do with now-discredited medical research programs directed at African Americans. The most notorious of these was the Tuskegee syphilis experiment, a long-term study on the clinical progression of untreated syphilis among African Americans named after the Arkansas district where it took place. Begun in 1932 and not abandoned until 1972, the Tuskegee experiment offered various health-care benefits to participants in exchange for regular checkups and blood tests. However, one benefit that was not made available, and was in fact deliberately withheld, was treatment with penicillin, whose ability to cure syphilis was known as early as 1947. Doctors keen not to disrupt the long-term aims of the study did not offer a treatment that would have helped their patients. Many died who might have lived. When the experiment was finally exposed in the *Washington Star* and the *New York Times*, it was immediately terminated. In the aftermath, clinical trial and consent protocols were overhauled to prevent a recurrence, and, in 1997, President Clinton issued an apology on behalf of the U.S. government. Even now, the memory of Tuskegee makes African Americans very wary of the intentions of official health programs and suspicious about the potential for some sort of eugenic misuse.

I became aware of this danger myself when advising a UK Parliamentary Select Committee inquiry into the impact of genetics. In my position as special adviser it was part of my job to recommend particular fields that the inquiry might cover and one of these was the effect of prenatal screening programs for serious genetic diseases. As well as considering the usual suspects like cystic fibrosis and muscular dystrophy, I also suggested we take a look at sickle-cell anemia and another inherited disorder of red blood cells called thalassemia. Both these diseases offer some protection to carriers against malarial infection, which is why they are both common where malaria is endemic. Thalassemia is found in the Mediterranean islands of Cyprus and Sardinia and is a health problem in those parts of Britain, like some London boroughs, with high densities of Cypriot immigrants. As we saw earlier, the mutations persist even where malaria is absent. In another superb example of well-organized genetic screening, like Tay-Sachs among Ashkenazi Jews, thalassemia has been virtually eliminated in both Cyprus and Sardinia.

The situation with sickle-cell disease is scientifically similar, but has again exposed the suspicions of the black community. Sickle-cell anemia is very common in West Africa, and is therefore a health issue not only in Africa itself but also in countries, like the United States and Britain, where West Africans have settled, voluntarily or otherwise. The Select Committee had invited a spokeswoman from the National Sickle Cell Program to one of the regular evidence sessions held in the Gothic surroundings of the House of Commons in London overlooking the Members' Terrace and the River Thames. In these sessions, each witness is interrogated by one member of Parliament with others joining in to ask supplementary questions. These are open meetings with members of the press and the public present, and all the evidence is published. From the dozens of distinguished witnesses and the hundreds of other submissions, I remember this particular session and the evidence from Marion McTair very well. I have a copy of the report on my bookshelves that contains a verbatim transcript of the evidence to the inquiry from McTair on behalf of the National Sickle Cell Program. The screening program was basically designed to replicate the very successful thalassemia and Tay-Sachs efforts to diagnose carriers and offer couples at risk of having a sickle-cell child some genetic counseling and also advice on the option of selective termination of affected pregnancies.

When Ms. McTair, who is black, was asked by one of the MPs for her opinion of the screening program, she recalled a visit from a lady who was a sickle-cell carrier herself and was also married to another carrier. The rules of genetics mean, with her and her husband both being carriers, that she has a one-in-four chance of bearing a child with sickle-cell disease at every pregnancy. After their first child, who was affected, she became pregnant again and decided on a prenatal DNA diagnosis. This unfortunately identified the fetus as having two copies of the mutated gene and thus bound to develop sickle-cell disease. Reluctantly she agreed to a termination. Again she became pregnant, and again, very unluckily, the prenatal DNA test came back positive for sickle-cell. At this point the woman went to see Ms. McTair to ask her advice about what to do. She said, "I don't want to abort it, I don't trust the doctors." "Why don't you trust them?" "Because I can't understand why this is the third time I'm carrying a child with sickle-cell disease. I think maybe the doctors don't want me to have this child. It's the wrong color, and it's not the type of child they want in Britain today. They want white ones, bright ones and not black, daft ones." Ms. McTair asked her why she said that and she could not say, but in her mind it was how she felt. What a contrast to the Ashkenazi Jews, who saw how genetic testing could help rid them of the scourge of Tay-Sachs and also to the Sicillians, who overcame their deep religious opposition to abortion, with the Catholic Church turning a blind eye, in order to achieve a comparable result and eliminate thalassemia from their island.

One in twelve African Americans carries the sickle-cell gene, which makes for about two million carriers. One in every 144 (i.e, 12 × 12) couples will both be carriers with a 25 percent chance of having a seriously ill child, so there is every reason to develop a screening program. But when the parliamentary committee came over to America for a week of evidence gathering, we heard that suspicion of a racially targeted eugenic motive lying behind a sickle-cell screening program in the United States had led to its suspension. Unlike the genetically parallel problem of Tay-Sachs disease among the Ashkenazim, where there have been no affected births for several years, even now, one in five hundred black babies in the United States is born with sickle-cell disease. This is a tragedy caused solely by the historical suspicion of the black community and one that doesn't appear to be getting any better, as it has now become enmeshed in political arguments about the "slavery health deficit" and calls for reparations. Paige has to deal with this innate suspicion all the time, which is one reason why, like Oxford Ancestors, her company destroys its customers' DNA samples as soon as the analysis is finished.

Paige is also in the front line of the sometimes angry reaction from African American men whose tests reveal that they carry a European Y chromosome, which roughly a third of them do. She knows very well that this is an emotional rather than an intellectual response, as most of her customers realize it is a possibility. And as with most emotional responses to unexpected news, things calm down after a while as people begin to accept the situation. After Paige has talked them through the initial stages of apprehension, they usually come to accept how it might be possible, especially when questions to other family members uncover half-forgotten tales of a white ancestor. This is not true for the reverse situation, as Paige explained with an example. A white man from Texas, with all his ancestors on both sides of the family white for as long as anyone knew, complained bitterly when his mitochondrial DNA came back as African. After the usual assertion that "there must be some mistake" and demands for retesting—which returned the same result—he still could not accept that he had a black ancestor. Of course this begs the question of why he chose to have his DNA analyzed by Paige's company, African Ancestry, in the first place.

We finished our conversation by talking about the future of the genetic genealogy business. All "long-established" organizations like African Ancestry, Family Tree DNA, and Oxford Ancestors have felt the impact of new companies offering much cheaper rates. These are usually no more than agents who act as intermediaries between their customers and a lab but know nothing about the underlying science or how to interpret genetic results. All three of us—Bennett, Gina, and ourselves at Oxford Ancestors—get frequent calls from customers who have bought on price from a cheaper outfit and then want someone to explain what their results mean. In genetics, as in life, you get what you pay for.

We left Paige at the station and in the gathering dusk, walked back past Capitol Hill and along the Mall to our hotel. Halloween parties were breaking out all over the place. You take it far more seriously in the United States than we do in Britain.

Paige had very kindly put us in touch with two of her customers with strong views on genetic testing among the African American community, and the next day we met up with them. The first was Toby Cooper, a bubbly and engaging woman who ran her own government contract business. The location was our hotel lobby, but this was an arranged meeting rather than a piece of DNA hustling. Toby already knew about her own mitochondrial DNA through African Ancestry, which had located her matrilineal roots to Cameroon in central West Africa. Her father's Y chromosome had its closest matches on the slave island of Bioko off the Cameroon coast. She was a great fan of this kind of testing and the solid links it had made to her African roots. After I had explained chromosome painting, she was equally keen to have a go at that. "I can't wait. Isn't it amazing what you can find out about yourself these days," she effervesced. She went to the ladies' room to provide the saliva sample, and when she returned with it safely in the tube, I asked her more about herself.

She had grown up in Aurora, Colorado, where there were very few blacks, and it was only when she moved to Washington, D.C., to complete her MBA that she became fully aware of the complexities and contradictions of being an African American woman. I was half expecting to hear her experiences of racial discrimination at the hands of whites, but instead she reserved her anger and frustration for the behavior of African Americans themselves. This began to be clear when she told us about a male friend, another African American, who said that he did not like her beautiful head of curly black hair, which she wore in a bob. He told her that he was embarrassed to take her to dinner at an upscale restaurant unless she blow-dried it straight. It was fine to be seen with her and her natural hair at a basketball game, but in his mind the elevated social circles to which he aspired demanded a more Eurocentric look.

She soon told him to get lost. This was the hair that had protected her ancestors against the tropical sun. That was why it was curly and dense in the first place, a natural helmet against the searing solar radiation. Looking at Ulla with her blond Scandinavian locks, she quite rightly said that fine and straight European hair would have been no good under the African sun. It would have let the damaging rays straight through and toasted the scalp. She was proud that she had inherited the hair of her ancestors, ancestors who had built the great civilizations of Africa. Cooper despaired of the multitude of black women she knew who, although they did not have enough for a computer or even for proper food, nevertheless found the money for long straight hair extensions every month, because they thought it made them prettier, more acceptable. Not to whites, she emphasized, but to other black men. This, Toby said, was very prevalent in Washington, where shops had swatches of long Asian hair hanging on the wall and more than a few black women buying them. They would staple, glue, and tape the hair to their heads, even if this meant they ended up with scalp infections and went bald. In Toby's opinion, they had been brainwashed into thinking, "the more white-like I am, the better I am." I was amazed by what Cooper was telling me, but then this was the first time I had ever had a conversation anything like this with a black woman. It turned out that this was also the first time she had ever discussed these things with a white man.

Toby knew from the work of African Ancestry that a third of African American men carry a European Y chromosome, and she had very strong views of how this came about. It was not just that European men had used their position as slave owners to force themselves on the women in their charge—though that would have accounted for a great number of the Y chromosomes but also that these couplings were at times part of a sexual bargain manipulated by black women themselves. According to Cooper, even before they were put on the slave ships, black women had figured out that if they became pregnant by a European man, there was less chance of being transported. Once in America, she told us, this particular type of relationship between black women and white men continued to flourish under slavery. "We knew," Cooper confided as if suddenly talking for a bygone generation of ancestors, "that if we could soften up the overseer or even let them sleep with us, they might not beat our husbands or our sons." Although there were no formal marriages between black slaves, I knew what she meant by "our husbands." In this way black women had made sacrifices to protect their men. "We knew how to manipulate and maneuver and use sex on behalf of our race," she continued. This did not end with the abolition of slavery, Cooper added, as white households were happy to allow black maids into the house, but not black men.

While black women could get domestic work, their men either toiled as low-paid sharecroppers or had no work at all. It was the women who supported their families through the hard times of the Reconstruction after the end of the Civil War, and it made Cooper angry that this seems to have been forgotten when today's black men try to get their girlfriends to make themselves look more European. Working as a maid brought with it opportunites for sexual contact between white and black, whether consensual or forced. That, in her view, was the explanation for the abundance of European DNA in African Americans.

Where sex is involved the potential for hypocricy is enormous, as Cooper reminded me when she told the story of Strom Thurmond. Senator Thurmond had died in 2003, aged one hundred, the only senator in the history of the United States to reach that age while still in office. He was vehemently opposed to racial mixing and a savage opponent of the Civil Rights Bill of 1967, conducting the longest-ever solo filibuster in the history of the Senate, lasting over twenty-four hours. Only after his death was it revealed that when he was twenty-two, he and his family's sixteen-year-old black maid, Carrie Butler, had a daughter, Ellie May. Thurmond never publicly acknowledged her, though he did pay for her college education among other things. After his death, his family acknowledged Essie Mae, making her eligible for membership of the exclusive organization Daughters of the American Revolution through her lineage descent from Thurmond.

Our interview had lasted just thirty minutes, cut short because I was due at a radio studio, but I learned more about African American women in that half hour than I had in my whole life. I was very glad indeed that Gina Paige had put me in touch with Toby Cooper, and very grateful to Cooper for being so frank and open. As she got up to leave, she handed me a copy of *White Women*, the vivid account of her observations and experiences that she wrote in 1998 and which lays them out even more forcefully than she had in our interview.[1]

After Toby left, we got straight into a taxi and headed for the studios on the other side of town. Gone were the wide boulevards near our hotel, a stone's throw from the White House. The streets were narrower and the buildings crowded in on us as the taxi weaved in and out of the traffic. By now it was dark and the people on the street hard to see. Any sense of unease was soon dispelled by the bright lights of the Sirius XM building and, once inside, we were ushered up to Mark Thompson's studio. I had gotten to know Thompson, though only briefly, through a live radio show for the BBC World Service in which he, Gina Paige, and I had taken part a few months before. It was a discussion and phone-in called *Africa Have Your Say*, and the topic was the familiar one of the impact of DNA on reinforcing African identity. I didn't think it was a particularly good show; it suffered from poor participant selection, so too much time was taken up by rants and irrelevances from the phone-in contributors. On the topic of African Americans who visited their tribal homelands, as identified through DNA, I was surprised to hear what an evidently agitated native African caller had to say. In strident tones he told the world that he was fed up with wealthy Americans coming over to his village and embracing their "ancestors," then hightailing it back to the airport as fast as they could. When asked by the flummoxed adjudicator what he would have liked from his African American cousins, his reply was both brief and direct. "A green card."

Thompson had impressed me by his evident strength of feeling about what his own DNA connection had meant. In an echo of Oprah Winfrey's famous declaration, Mark answered the increasingly bewildered adjudicator's question of what he felt about his own identity after the DNA test. "I am African," he retorted without a moment's hesitation. The slot on his show went well enough, I thought, but there were a lot more things I wanted to ask than there was time for. Besides, it was his show, so he was asking the questions. Very generously for such a busy man, Thompson agreed to come around to our hotel for breakfast the next day.

Thompson arrived and joined Ulla and me around the table. He told me that his father and mother had met in Washington when she was a student of fashion design. Thompson was born at the Freedmen's Hospital, which building now houses Howard University's School of Communications, a coincidence not lost on the broadcaster. He and his mother moved back to her family in Nashville, Tennessee. His father was meant to follow but he did not, so his maternal uncle, his mother's brother, took on that role, a custom that Thompson later discovered was almost the norm in Africa. Nashville is a college town with Fisk University and Meharry Medical College (Roy King's alma mater) being among the first universities, like Howard in Washington, specifically created for black students. Thompson grew up in faculty condos on the campus where, for more than thirty years, his mother was on the administrative staff. This was the early seventies, when black cultural and intellectual activism was reaching a zenith after the civil rights legislation of the late sixties. Everybody involved in the civil rights movement passed through Nashville at one time or another, and the young Thompson was there to see them in the flesh. This infused him with a strong sense of African American culture and of black activism that has stayed with him ever since.

At eighteen Thompson moved back to Washington as a student at Georgetown University, and it was not long before he clashed with the authorities. This was the time when the tide of international opinion was moving against apartheid in South Africa and there was pressure on the U.S. and British governments to impose economic sanctions. Ronald Reagan was president at the time, with Margaret Thatcher his opposite number in Britain. They were both against sanctions, advocating instead the policy euphemistically referred to as "constructive engagement." Their refusal to condemn apartheid, or at least not to do much about it, sparked a wave of popular protests on both sides of the Atlantic. Thompson got involved when the student body at Georgetown put pressure on the university to pull out from its substantial investments in South Africa. The crunch came when Reagan's secretary of state for Africa, Chester Crocker, arrived in Georgetown to give a speech, only to be met by crowds of students chanting, "Free Nelson Mandela." Thompson covered Crocker's speech for the student newspaper, and his account grew into an in-depth article criticizing U.S. foreign policy toward South Africa, an article that the State Department asked the university not to run, which only served to fan the flames of protest.

Within a few weeks the Georgetown students took over the university's School of Arts and Sciences, built a South African–style shanty on the campus, shut down the schools, and forced the administration to disinvest. They got their way, but they all went to jail for a spell. As we all know, Nelson Mandela was freed and became the president of his country, going on to become one of the world's most venerated public figures. But it was not always like that. As the final twist in Thompson's account of his days as a student activist, he told me that he met Mandela when he visited Washington and that the meeting took place in this very hotel, not ten feet away from where we were sitting having breakfast.

Thompson had already started broadcasting on a black talk station when he got his own show on Sirius XM Radio in Washington in

2001. Not long after that he met Gina Paige and began working closely with her because he felt very strongly that African Americans needed to reestablish their links to Africa, not only for themselves but also for the people of Africa. He had visited the notorious Elmina Castle on the coast of Ghana and found the experience both moving and troubling. In one part of the castle, he told me, is the "door of no return" that leads from the dungeons where the slaves were held and opened straight out onto a ramp down to the shore. Once through that door, there was only one way to go, onto the waiting ship and across the Atlantic to the New World. On Thompson's visit he and the others in his group had talked about what would have happened if, instead of going to the New World as slaves, their ancestors had stayed where they were. At this point a group of children gathered around offering candies for sale. Thompson asked one of them, a girl of about fourteen, what was her name. "Elizabeth," she replied. That was Thompson's mother's name, his grandmother's, and that of his female ancestors for as many generations back as he knew. The same was true of the small girl, her mother and grandmother were both Elizabeth. It was she, not Thompson, who then said that perhaps, if they were able to go back far enough, they would find that they were both descended from the same ancestor.

It was very clear to me that Mark Thompson senses the feeling of a severed connection with Africa and that one of his ambitions in life is to help African Americans to regain it, which is the reason he supports and works so closely with Gina Paige and African Ancestry. He has become fascinated by the link between science and spirituality, especially since he became ordained as a minister, and he understands the special qualities of DNA. I now asked him the key question. All through his career as a political activist he had concentrated on his African roots. How would he feel if the DNA test that we were about to do showed that he had other ancestors too, maybe Native American or Asian or, more likely, European? Thompson told me that his mother was very pale-skinned for an African American, but she was raised black. As he was growing up as a kid, people would ask him whether his mother was white, although it was always clear from the way she spoke and carried herself that she was a black woman. It isn't something that many black people want to acknowledge, but given how fair his mother was, he thought he probably did have some European in his DNA. His father was much darker skinned, and his own pigmentation, he told me, was somewhere in between.

Thompson then went on to tell me that he was doing a little family history research of his own. His maternal grandmother's surname was Polk, and her family had lived in rural Tennessee since at least the 1800s. James Knox Polk was the eleventh president of the United States, and his family had a plantation in the same part of Tennessee. During his White House years, between 1845 and 1849, it was Polk's brother William who looked after the plantation. Thompson's family traced back to a William James Polk, living at about the same time. Though Thompson knew that it was commonplace for slaves to take on the surname of the plantation owner, there was a distinct possibility that, rather like the descendants of Thomas Jefferson and Sally Hemings, Mark might also have a direct line back to a president of the United States. Any European DNA that we found might possibly have that special pedigree.

I was very impressed by Thompson and by his thoughtful devotion to his African roots. He had discovered through Gina Paige's DNA test that his Y chromosome points to an ancestral origin in Sierra Leone while his mitochondrial DNA has come down a long line of matrilineal ancestors from further south, in Cameroon. Before that, like many African Americans, he had cemented his ties to Africa by adopting an African name, Matsimela Mapfumo, which is a blend of South African and Zimbabwean meaning "Firmly Rooted Soldier." When his DNA results told him that his African ancestry, at least for the two lines tested, was from West Africa rather than the South, the news, if anything, strengthened his feeling of connection. As he explained, when you adopt an African name you choose where you would like to have come from, but when you find out through DNA, then you know. Through his show, the activist in him was beginning to challenge African Americans to become involved and accept some responsibility for the part of Africa that Paige's DNA tests had identified as their ancestral homes. As we spoke, he was trying very hard to find African Americans whose genetic lines went back to the Congo, the country that, in his view, was in the biggest trouble. He despaired of the minimal coverage given to Africa by the famously insular American media and said he relied on the BBC for most of his information. This was one legacy of British colonialism that he approved of, he said with a wry smile as we left.

In the planning stage for *DNA USA* I had come across a reference to a working group on genetics and genealogy organized by the W.E.B. Du Bois Institute of African and African American Studies at Harvard. The group had been convened by the institute's director, Henry Louis Gates Jr., and it looked both exciting and relevant, so I made a mental note to get to see him if I could arrange it. Before I had done anything about it and completely out of the blue, I got a message from a television production company in New York asking whether I would be willing to participate in a documentary with the same Dr. Gates. Of course I was delighted and accepted immediately. Another piece of serendipity. "If you build it they will come."

The chosen location for this televised encounter was the American Museum of Natural History in New York, and the date proposed fitted perfectly with the end stages of the journey back to Boston for the return flight to England. Ulla and I arrived at the museum first and waited in the lobby by the dinosaur skeletons. I looked outside to see a film crew gathering at the foot of the steps. A car pulled up, and out stepped Dr. Gates. He was elegantly dressed in a dark blue suit, a beautifully pressed white shirt with a maroon tie, and a matching handkerchief neatly folded in his jacket pocket. In his left hand he carried a cane. In complete contrast, having spent three months living out of a suitcase, both I and my outfit could be summed up in a single word: crumpled.

Since filming could not begin until after the doors had closed, the museum had been hired for the evening, and we took over the lower ground floor with its charmingly old-fashioned dioramas of the African plains and the jungles of Borneo, exhibits of a sort that have sadly disappeared from most modern museums. The filming went pretty well as far as I could tell, with each of us kidding the other as we talked about why it was that mitochondrial DNA and Y chromosomes fell into distinct clusters all started off by one person. We filmed late into the night, after which Dr. Gates gave Ulla and me a lift back to our hotel. There was no time for a long conversation in New York, so we arranged something for a few days hence.

As it was, it would be several months before we met again. This was when Ulla and I returned to Boston to meet up with the New England volunteers to go over their DNA results, which, by that time, had come back from the lab. I had timed our visit to coincide with the Annual Dinner of the New England Historical and Genealogy Society, where the main speaker was to be Annette Gordon-Reed, whose biography of the Hemings family and their relationships with Thomas Jefferson I had read and admired. Dr. Gates had also been invited to the dinner as a guest, to be honored with a specially commissioned genealogy by the society. I took this opportunity to arrange to visit him in Harvard a couple of days later with the promise of an interview.

Between our meeting in New York and the approaching rendezvous in Boston, I had read up on Dr. Gates and soon realized quite what a prominent and highly regarded individual he was. He is foremost a scholar, the author of fourteen books so far, including *The Signifying Monkey*, a theory of literary criticism that went on to win the American Book Award, and *The Trials of Phillis Wheatley*, an appraisal of America's first published black poet. However, he has always had a great personal interest in genealogy, and his highly acclaimed television documentary series introduced him to an audience far beyond the narrow confines of academe.

Gates could not have been more accommodating when Ulla and I met him at the Du Bois Institute in Harvard. He took us on a whistle-stop tour of the institute, where we saw the world's only hip-hop archive, with recordings, memorabilia, and even the footwear of famous rap stars, and, in separate libraries, a fabulous collection of African and African American art, including one project titled *The Image of the Black in Western Art*. I met his staff, again extremely helpful, and all clearly devoted to Gates. Next we dashed to his house to check on his own DNA results. He had already had his chromosome portrait painted and also his mitochondrial DNA and Y chromosomes analyzed. There wasn't time for me to look in detail, but I did immediately recognize from the mutations in his mitochondrial DNA that we were both members of the same clan and closely related through our maternal ancestors. No wonder we were getting on well. To Ulla and me he was completely charming, open and extremely helpful. He had enjoyed *The Seven Daughters of Eve* and, in a very gracious gesture, offered to put my name forward for a fellowship at the institute, an offer that I gratefully accepted.

Gates is well known across the United States as the host of the successful PBS documentary series *African American Lives* (2006), *African American Lives 2* (2008), and *Faces of America* (2010) in which I made a brief appearance. In these shows Gates introduces a number of prominent Americans to their own ancestry. He was quick to realize the special place that DNA can play in connecting African Americans to Africa and injected that element into all three programs. The genetic component of the most recent program had been helped a great deal by the proximity at Harvard to the Broad Institute, a high-throughput DNA research facility founded by two luminaries of genetics, Eric Lander and David Baltimore, which is devoted to untangling the genetics of common disease. I had known these two distinguished scientists as champions of the human genome and also a third, George Church, a professor of genetics at Harvard Medical School who was affiliated with the Broad. Through his membership of Church's Personal Genome Project, Gateses junior and senior were the first father and son, and the first African Americans, to have their entire genomes sequenced. Since then, the Du Bois Institute has gone on to cosponsor the Roots into the Future initiative, a large-scale program of genetic testing, directed specifically at African Americans, that is searching for sophisticated correlations between genetics and treatment in this community that has, thus far, been distressingly under-represented in the genetics data.

In *African American Lives* Gates employs mitochondrial and Y-chromosome tests on his guests in a way that is familiar to us. He also used the AIMs markers to estimate the ethnic composition of their genomes. Remember that when these documentaries were made, in 2006 and 2008, chromosome painting was not available. Even so, I was keen to ask Gates about himself, his experiences making the shows, how his guests had reacted, and lots of other things. We met over dinner with some of his colleagues from the faculty. The location was the Charles Hotel on Harvard Square, and the diners, other than Dr. Gates, Ulla, and me, were Marcyliena Morgan, an anthropologist, a professor of African and African American Studies, and the director of the hip-hop archive; Lawrence D. Bobo, W. E. B. Du Bois Professor of the Social Sciences at Harvard and a member of the prestigious National Academy of Sciences; and Evelyn Brooks Higginbotham, a professor of History and chair of the Department of African and African American Studies. On paper, this was a high-class academic gathering with more degrees around the table than pieces of cutlery. However, it was no dry encounter but one where serious discussion was punctuated by witty repartee and loud peals of laughter. Ulla and I have rarely had such an entertaining evening.

I had prepared a list of questions in my notebook. The first of these referred to Gates's own DNA. As mentioned a short while back, he and I are closely related through our mitochondrial DNA. That is not because I have an African matrilineal ancestor, but because Gates is a member of one of the seven European clans and shares my clan mother, Tara. It is quite unusual for an African American to have a European mitochondrial DNA, but that was not the point of my question. Rather, it concerned the way this news was relayed to him by the lab. One of the first things people want to know when they get their DNA results, naturally enough, is who they match and where their newfound relatives are from. When Gates had asked this question, he was told he had a match in Nubia, a region of Sudan on the River Nile to the south of Egypt.

What he was not told was that he also had hundreds of other matches in Europe, which is no surprise for a descendant of Tara. I was interested in the reason for this omission. When Gates saw a map of the Old World with his mitochondrial DNA matches marked on it, sure enough there was one red spot in Nubia, but to the north and west, the spots were so dense that, in his own words, "it looked as if Europe had smallpox." So why was he not told about these genetic matches in Europe? When he challenged the lab, not the Broad on this occasion, they said that they thought he might be disappointed. "Were you?" I asked. "No, not at all," he replied, but it still struck me as interesting that the lab thought he might have been.

We then moved on to how African Americans reacted to discovering the extent of their European ancestry. It is very unusual for them to have non-African mitochondrial DNA, but very common for African American men to carry European Y chromosomes, which roughly a third of them do, as we have seen. In my conversations with Gina Paige and Rick Kittles, they were very aware of the potential for disappointment that people felt when confronted with this particular piece of news. That chimed with my much more limited experience when I had been asked to do the same test on some Afro-Caribbean celebrities in Britain for a BBC documentary, *Who Do You Think You Are?* In one case I remember, a top-class athlete had an obviously European Y chromosome, but the producers kept that from him and it was never shown. Yet Dr. Gates's view was that it really didn't matter all that much and that African Americans would, or should, be very comfortable with discovering that some of their ancestors had European roots. Behind his answer was, I think, one of the founding principles of the Du Bois Institute that is so central to Gates's vision: African American culture was not merely African across the Atlantic but a separate culture, worthy as a subject of scholarship in its own right, irrespective of the identity of the scholar or student. In other words, you didn't need to be an African American to study the culture any more than you had to be English to appreciate Shakespeare.

However much African Americans know about the general conditions of their ancestry, Gates found—as Rick Kittles had done, and as I had done on a much smaller scale—that they can sometimes be shocked when they first hear that so much of their DNA is rooted in Europe. We have already seen the sequence of reactions from initial surprise or even anger to eventual acceptance among the customers of African Ancestry. But there are some African Americans who are sure that absolutely all their ancestors are from Africa. Probably the most famous of these is Oprah Winfrey. She is also numbered among Dr. Gates's friends and was one of the guests on the first series of *African American Lives* and the subject of his book *Finding Oprah's Roots.* For this she took a mitochondrial DNA test the day before she was due to leave for South Africa to visit the orphanage that she was sponsoring. As soon as she arrived in South Africa, she made her famous quote "I am a Zulu", which was flashed around the world, much to Gates's surprise. As the golden rule of DNA television is to keep the results a secret until the on-camera "reveal," he immediately rang the lab to ask if they had given the results to Oprah. They answered that they certainly had not, and, in any event, as the DNA sample had only just arrived, it would be several days before they had the results. When the lab results did come back, it was obvious that Oprah's mitochondrial DNA matched others from West Africa, in particular Liberia, much more closely than anyone from Zululand. That was always the more likely origin. Zululand, which is situated in South Africa far from the coast, is a long way from the main centers of slavery. Showing a certain amount of surprise but certainly not indignation at having her roots moved halfway across the continent, Oprah quickly and graciously assimilated the new information. She may not have been a Zulu, but she was certainly African, a point reinforced when an AIMs test found no trace of European DNA. A remarkable woman in more ways than one.

At this point I slipped my own DNA into the conversation. From a genetic point of view, I was part African myself. My own chromosome portrait had revealed that a DNA segment at the end of the short arm of my chromosome 11 was half European and half African. "If I were an

American, would I qualify as an African American?" I inquired. Bobo, who laughs even more zestily than Gates, said that the infamous "one-drop" rule would almost certainly have classified me as a slave in the old days. The short arm of chromosome 11 is packed with important genes, one of which controls insulin production. Thus one of my two copies of the insulin gene is African, and since the only place insulin is made in the body is inside the pancreas, I considered myself, metaphorically at least, to have a black pancreas.

As the evening progressed, we meandered through a wide range of topics: The situation in Brazil, where there are an astonishing 134 unofficial racial categories. The case of P.B.S. Pinchback, the first black acting governor of any state, Louisiana, who was sworn in during 1872 and who, although the son of a slave, appeared very white indeed. "You would look like Michael Jordan next to him," quipped Gates. The sad case of the writer and critic Anatole Broyard, whose daughter Bliss wrote the touching biography *One Drop*. Broyard was diagnosed with prostate cancer and, because of his appearance, treated as a white man. Had his doctors been aware of his true ancestry, which was partly black, they may have treated his cancer more aggressively and in so doing perhaps saved his life.

As the party eventually broke up, I felt so very fortunate that my profession had brought me into contact with such wonderful company. The long reach of genetics, far beyond the confines of the laboratory bench, had led me all over America and the world. Neither at supper that night, nor in any of the revealing conversations that I had been privileged to have with African Americans during my journey, had we gotten anywhere near "solving" the relationship between genetics and race. There is no "solution." But knowing that at least one of my ancestors had been black, and that my own pancreas depended on that legacy, had certainly blurred the edges of my perception and, not for the first time, brought home to me the myriad threads that bind us all into the same human family.

THIRD MOVEMENT

19

The Private View

Scientific papers traditionally end with a "Discussion" section, and the temptation here is to follow that convention and write a final chapter that summarizes the rest of the book and comes up with ponderous conclusions. But that is not how road movies end, is it? Most of the time the protagonists never get to their destination and end up, like Peter Fonda and Dennis Hopper in *Easy Rider*, blasted by a shotgun and dead at the side of the road. Even if I was feeling the worse for wear at the end of three months on the move, my chest was only wheezy, not full of lead shot.

I did return with Ulla to the United States six months later, carrying with me an armful of chromosome portraits and ready to go through them with my volunteers. Time and money confined the face-to-face meetings to the east coast, so I sent individual chromosome portraits to everyone else by snail- or e-mail, and followed up with a phone call. I had asked the volunteers to make a stab at what they thought their own portrait might look like before I showed it to them, and then observed how they reacted when the canvas was revealed. The journalist in me was waiting for a series of dramatic outbursts when expectations met reality, but there were none. Most of the time my volunteers either declined to hazard a guess or, when they did, found that it more or less matched what their portraits revealed. When it did not, the reactions were muted and reflective rather than indignant. But what do the portraits look like? I will let them speak for themselves.

Welcome to the "Private View." Take a glass of champagne in one hand and a canapé in the other. Think of this last chapter as the catalog. Go to the color insert section and take your time to browse.

The first portrait in the gallery is of my New England volunteer "Margo Channing." She is, as you see, all one color: blue. This means that without a single exception, all of "Margo"'s DNA has a European origin. It is not the most colorful portrait or, you might think, the best to choose to begin our tour of the gallery. More Rothko than Magritte. But, despite its uniformity, this portrait is one of the most interesting and surprising of all. I could equally well have shown you the portraits of my other New England volunteers: "Terry Malloy," "Rio McDonald," "Anna Christie," "Lisa Fremont," or "Rose Sayer."* All of them are exactly the same, solidly blue throughout without a trace of DNA from either African or Native American ancestors. Remember that all my sitters in this part of the gallery are descended from European settlers who arrived in New England during the seventeenth century, many of them before 1650. And yet there is not the slightest echo of any interbreeding with Native Americans, with whom they lived in close proximity. Had there been, then the blue chromosomes would have been flecked with orange. That was what I had been expecting, having seen the effects of European settlement among other indigenous people where genes are quick to cross ethnic boundaries in both directions. But in New England there is no sign of it.

I can only conclude that, if there had been any intermixing between early New Englanders and the indigenous tribes living around Plymouth and Cape Cod, the offspring would have stayed within the Indian tribes

---

* Quotation marks have been used to denote pseudonyms.

rather than being absorbed into the English settlements. The single exception is "Atticus Finch." His portrait shows a tiny fleck of orange at the end of chromosome 9.

"Atticus," if you recall, had reason to believe he was descended on his father's side from Ots-Toch, the daughter of a Mohawk mother and a French father. Although I cannot prove it beyond any doubt, I think it likely that the speck of orange in "Atticus"'s chromosome portrait really is the genetic legacy that he inherited directly from Ots-Toch and her Mohawk mother. The fact that the portraits of my other New England volunteers were all uniformly European blue throughout makes me think that the little bit of orange in "Atticus Finch"'s portrait is genuinely from his Mohawk ancestor. After twelve generations of doubling dilution, there was never going to be much left of her DNA in "Atticus"'s genome, and the single speck we see in his portrait is about all I would have expected.

One of the attractions that drew me to the chromosome portraits, as well as their visual impact, is that it is easy to look up which genes correspond to which particular chromosome segments. The Human Genome Project located all our genes at fixed points along the chromosomes, so it is a simple task in "Atticus Finch"'s case to identify which genes have come down to him from Ots-Toch. This will identify the parts of "Atticus"'s body that are running on Mohawk DNA. Among several genes that chromosome 9 carries with frankly obscure functions, there is one that is familiar to all of us. This is the gene that controls our blood group, deciding whether we are group A, B, AB or O. We all have two copies of this gene, one from each parent, and in "Atticus"'s case one of them has been inherited from Ots-Toch, while the other has come from a European ancestor on his mother's side of the family. Both Mohawk and European genes are working together to decide "Atticus"'s blood group, and since he is fit and well, it looks as though they are doing a good job.

"Harry Lime," a staff member at the New England Historic Genealogy Society who also volunteered to have his chromosomes painted, was the only other European New Englander to have anything other than a uniformly blue chromosome portrait. In his case, as you can see, the speck of color was not orange but green. "Harry Lime" therefore has an African ancestor. When I unveiled the portrait, his first question—to himself more than to me—was to wonder who this ancestor might have been. Being a professional genealogist, he set off in search of this unexpected family member. Like "Atticus Finch"'s portrait, this was only a splash of color amid a sea of blue, suggesting that, whoever it was, his African ancestor had lived a very long time ago. The likelihood is that he or she was an African American, but even that is not certain. Some Europeans, myself included, have small segments of African DNA that must have entered the British gene pool at some time in the past. The last time I spoke to "Harry," he was still on the track of his elusive African ancestor.

As we did with "Atticus Finch," we can also see which of "Harry"'s genes are firing on African DNA. The section of chromosome 7 containing African DNA does not hold any well-known genes. However, there is one intriguing gene among the otherwise uninspiring collection. It belongs to the family of genes that control the exquisitely sensitive receptors that give us our sense of smell. It is a large family dispersed around the human genome, with each member of the family capable of sensing particular odors. In "Harry"'s case one of these is being run by a collaboration of DNA from one African and one European ancestor.

For proof that Britons can also have African DNA, we need look no further than my own chromosome portrait. As you can see, it has a small segment of green at the tip of chromosome 11. I don't know which ancestor this has come from, but black Africans have been coming to Britain since at least the time of the Roman Empire. Indeed there was an influx of African Americans who moved to Britain after the Revolutionary War, having been persuaded to fight on the British side with a promise of a guaranteed welcome in England after the war was over. In fact the promise was never kept, and most were shipped off to Nova Scotia, although some did eventually make their way to Britain. So, while my African ancestor probably came to Britain a very long time ago as a Roman slave, he or she might instead have been an African American, just as "Harry"'s African ancestor could have been British.

The particular segment of African chromosome 11 that I have inherited happens to be very rich in genes. Among many others, there are genes for insulin and for beta-globin, one of the two subunits of hemoglobin located there. So both my body's pancreatic insulin and hemoglobin output are controlled by a fifty-fifty mixture of African and European DNA. I also have a fleck of orange from an unknown Asian ancestor, and the segment of chromosome 7 that contains it houses an important collagen gene. So my skin and bones owe a great deal of their mechanical strength to my Asian ancestor, just as much as my pancreas is jointly run on African DNA.

Judging by her portrait, "Phyllis Dietrichson," like "Harry Lime," also has an African ancestor and although I met her through the NEHGS headquarters in Boston, some of her family are from North Carolina. Like "Harry," she does not know who this African ancestor was, but she was delighted when I told her that the chromosomal segment involved, which as you can see is located on her chromosome number 7, contains an important muscle gene called beta-actin. With this new knowledge, "Phyllis" now flexes her African biceps with increased vigor.

On our travels around America we met with "Rhett Butler," and, thanks to Ulla's powers of persuasion, before long we had a sample of his DNA. "Rhett" was a European American from the South, from Georgia, whose ancestors had come over from England in the early eighteenth century. He was quite sure he did not have any black ancestors, but when his portrait came back, there were three streaks of green against an otherwise all-blue genome. His reaction to this when I showed him was amusement more than anything else. By the sound of it, his sister had married a man who was a bit of a racist, and "Rhett" was looking forward to telling him that he had actually married a black woman, on the very reasonable assumption that at least some of his African ancestor's genes had also been inherited by his sister.

The portrait of another of our volunteers, "Sugar Kane," also shows up several long segments of African DNA. If you recall, Ulla and I met "Sugar" and her husband in San Francisco, where she had told us about the family photograph of her grandmother who looked to her like a Native American. It was because of the possibility of an Indian ancestor that "Sugar" had become very interested in the spiritual life of Native Americans and had gone on her own spirit quest, including a spell in the sweat lodge at Pine Ridge. "Sugar" and her family had lived in Florida for generations, and from the appearance of her chromosome portrait, she certainly does have one or more black ancestors. If you look closely you will see that she also has two very short smudges of orange from what was very probably a Native American ancestor. There isn't enough there to indicate that the grandmother in her family album was a full-blooded American Indian, but as we shall see, that is not always easy to tell from a genetic analysis.

Our other volunteer who, like "Sugar Kane," certainly considered herself to be from solidly European stock was "Ilsa Lund." Appropriately enough "Ilsa"'s family hails from the Carolinas, and her chromosome portrait—the one that gave Joanna Mountain such a shock when she first saw it—has even more green segments than "Sugar Kane"'s. As you see, the green segments in "Ilsa"'s portrait are quite long and relatively intact, which suggests they have been inherited from a fairly recent African ancestor. That is because there has been less time for the segments to be broken up by the constant shuffling that chromosomes undergo between each generation. However, despite having her cousin Greg on the case, the last time we spoke, "Ilsa" had been unable to identify any black ancestors in the family. Tellingly, she has not let her mother know. Like "Sugar Kane," "Ilsa"'s portrait also has the tiniest flecks of orange on four of her chromosomes, indicating a far distant Native American ancestry as well.

I am the first to point out that the gallery has far too small a number of individual portraits for me to draw any statistically significant conclusions. But that isn't going to stop me making a few observations on the collection for the catalog notes. I am genuinely astonished to find so little genetic evidence of intermixing among the descendants of the early New England settlers. Of the twelve complete genome scans, each one scrutinizing half a million markers, I found a segment of orange only in "Atticus Finch"'s portrait, and he probably inherited it from the Mohawk ancestor he knew about. Apart from the one segment of African DNA in "Harry Lime," every other New Englander has a completely European set of chromosomes. The volunteers themselves were not as surprised as I was, but I think that is only to be expected. After all, none of the European American volunteers from the South thought they had any African ancestors, yet they all did. The only explanation I can come up with for the completely blue portraits from New England is that if there were any liaisons with Native Americans, the offspring were never, or almost never, accepted into the English colonies and instead were raised as Indians.

Again with far too little to go on, the observation that all my white volunteers from the South did have at least some African genetic ancestry is in such stark contrast to New England that I think it is probably a genuine finding. Of course, the irony of the "one-drop" rule, which condemned anyone with the least bit of black ancestry to the life of a slave, is highlighted by the implication from these DNA results that many whites with deep roots in the South have some black ancestors.

The portraits of my European American volunteers had produced fascinating results, even if the portraits themselves were rather monotone, being mostly blue with the very occasional brushstroke of green and orange. This all changed when I unwrapped the portraits from my African American DNA sitters, which you will see as, catalog in hand, we make our way to the next room in the chromosome gallery.

In this section of the exhibition you can see portraits with a full range of color from the almost all-green Toby Cooper, through blue *on* green in "Virgil Tibbs"'s chromosomes, to the balanced blend of blue *and* green in the portrait of Mark Thompson. However, as you can see at a glance, unlike the New Englanders, not one of the portraits of my African American volunteers has a completely uniform set of chromosomes. Every single one shows the genetic evidence of at least some European ancestry. The other feature to notice in this section of the gallery is that all of the African American portraits also show some blocks of orange from Native American ancestors. There are some aspects of the brushwork that I think may indicate slight inaccuracies in assigning the Native American component against an otherwise African background. To be more precise, I am surprised that so many of the orange segments among African Americans extend across both chromosomes, which if taken literally would mean that they had been inherited from a common ancestor, which I find most unlikely. But this is only a detail and does not diminish the observation that all the chromosome portraits of my African American volunteers have some degree of Native American ancestry. This news delighted many of my volunteers who, like "Virgil Tibbs" from Boston, had hoped for a tangible genetic link to the indigenous inhabitants of their adopted land. Another notable feature of the collection is that the chromosome portraits of my African American sitters from the South, like Toby Cooper and the two friends from Atlanta, "Ned Land" and "Mildred Pierce," have a lot more green in them than the portraits of their fellow African Americans, like Mark Thompson and "Virgil Tibbs," whose ancestors had moved north either directly after the Civil War or later on in the nineteenth century.

Moving to the final section of the gallery, I have already explained why I did not want to paint the genetic portraits of Native Americans for *DNA USA*. However, I was fortunate to meet up with "Will Kane" and "Roger Thornhill," who each had one Native American and one European parent and who were gracious enough to give DNA samples. When I unwrapped the paintings prior to displaying them, I could see, as you can, that both portraits have a large component of blue from their European parents. However, the way the portraits are painted means that this contribution can easily be subtracted, leaving the lower half of each chromosome representing the Native American component. In both "Will"'s and "Roger"'s portraits you can see that these lower segments are a mixture of blue and orange. We have seen earlier in the book that although portraits of Native American chromosomes appear to have a

European component, this can be partly due to the Asia/Europe border artifacts of Siberian chromosomes rather than a genuinely recent European admixture. Mike MacPherson's estimate is that chromosome portraits of Native Americans living five hundred years ago, before there was any chance of interbreeding with recently arrived Europeans, would be 75 +/– 15 percent orange with the remainder blue. On these grounds the Navajo and Hopi ingredients of "Will Kane"'s and "Roger Thornhill"'s genome are well within the range for unmixed Native American chromosomes. What is noticeable is that there are no signs of green, meaning African ancestry, in either of the portraits.

"Will"'s Navajo and "Roger"'s Hopi chromosomes contrast dramatically with the final portrait in this room from my one and only Cherokee volunteer, "Lucas Jackson." I was astonished when I first saw his chromosome portrait, and so was he. "Isn't that something!" he said with quiet amazement. There is only one small segment of orange among an otherwise uniform sea of blue. I would have dismissed this as an error were it not for something Mike MacPherson said when I visited him in San Francisco. He had evidently had a similar experience with the company's Cherokee customers, and had often found very little sign of orange in their chromosome portraits. We did not discuss the "Cherokee paradox," as MacPherson called it, any more than that, but it did make me think that perhaps "Lucas Jackson"'s portrait was not so unusual for a Cherokee as I had first thought. It also shone an admittedly dim light on the question of tribal membership of the Cherokee Nation and the other displaced Oklahoma tribes. Though "Lucas" has not yet applied, his father and many of his relatives still living in Oklahoma are members of the Cherokee Nation through their proven genealogical descent from ancestors on the Dawes Rolls. Yet, even though one of his chromosomes came from his European mother, there is really only one tiny speck of Native American DNA on the chromosome he inherited from his Cherokee father, which is far less than the average of 6 percent shown by the disenfranchised descendants of the Freedmen of Oklahoma.

Coupled with the almost complete absence of any Native American paternal or maternal ancestry among the Seaconke Wampanoag, as revealed by the Genographic Project, this only goes to show how incompetent DNA really is at assigning individuals to discrete categories. That is not the same as saying that race or ethnicity have nothing at all to do with genetics. After all, it is true that most African Americans do carry more African DNA than European Americans, and Ashkenazi Jews are more likely to be members of the clan of Katrine than the average native European, for example, but the correspondence is far too weak in individual cases for accurate assignment. However, the real problem here is not the basic genetics, which only seeks to describe underlying patterns of biological inheritance, but the human desire to create categories in the first place. Not wanting to wander into the well-trampled territory of the social and political ramifications of race and ethnicity any further than I already have, let me just quote one of my heroes, Arthur Mourant. He was among the tireless pioneers of blood grouping in the 1950s, when biology was first being proposed as a way of dividing up the world into discreet population units. He realized the fallacy when he wrote: "Rather does a study of blood groups show a heterogeneity in the proudest nation and support the view that the races of the present day are but temporary integrations in the constant process of . . . mixing that marks the history of every living species."[1] Substitute DNA for "blood groups" and modernize the literary style, and this could have been written today.

There is only one more observation to make in *DNA USA*, and it derives from the fascinating complexity of the human genome illustrated so well by the chromosome portraits of the sitters. This is most readily appreciated in the multicolored paintings of my African American volunteers, because here you can see most clearly how all of our bodies work on the intimate collaboration of DNA segments handed down by thousands if not millions of ancestors. The Native and European American genomes are equally complicated mixtures of ancestral contributions, but it is not so easy to see the individual components in the present color scheme.

To illustrate my point I have picked out 140 genes that help run eleven major body systems and shown their locations against the chromosome portrait of Mark Thompson, one of my African American volunteers. The composite portrait is the last in the gallery and I have put the detailed description of these genes in an appendix. These are only a tiny fraction of the total number of genes we need to keep going, but enough to convey the principle. In all of us they work equally from the two copies that we inherited, one from each of our parents. They have to cooperate properly, or we would simply not survive, as the example of severe inherited disease teaches us. So, whatever their own individual ancestry, whether African, European, or Native American, our genes must have found a way of working together. My pancreas functions on a combination of both African and European genes. Equally, "Rhett Butler," a white man from the South, depends on the DNA inherited from an unknown African ancestor for his heart muscles to work properly. "Ilsa Lund"'s digestive system, and much else besides, runs on DNA from her African ancestors. "Atticus Finch" needs his Mohawk genes to make sure his red blood cells do their job well.

Unfortunately, in one respect, I did not get to paint the chromosome portrait of any self-confessed white supremacists or members of the Ku Klux Klan. Had they been from the South, then there would have been a very good chance that their portraits would have a few splashes of green. If that had been met with shock or denial, my next question would have been whether they would rather do without the African DNA. If the answer had been yes, I would point out that doing so would cause them to lose the use of their kidneys or heart muscles—or whatever the African DNA was doing for them. Would they go ahead and have these organs removed? I think not.

The essential genetic collaborations are even more obvious in the portraits of my African American volunteers. Mark Thompson's insulin output is controlled by 100 percent European genes at the tip of chromosome 11, making his pancreas less "African" than mine. The same goes for his beta-globin genes, located right next door. They are both from

European ancestors, which means that even though he is an African American he is extremely unlikely to be a carrier of sickle-cell anemia, which is caused by a mutation in this gene. I, on the other hand, could be a carrier, as this globin gene is located in the segment that I inherited from my African ancestor. This illustrates a relevant health-care issue. No doubt with the best of intentions, ethnicity is taken into account when deciding on diagnosis and treatment plans. Since I am easily classified as a white Caucasian, no one would ever suspect that I might be a carrier for sickle-cell anemia, which is an African disease, but I could well be.

During the Korean War in the early 1950s, American troops in the field were prescribed antimalarial drugs. About 10 percent of African American soldiers developed severe anemia after this treatment while European American soldiers only very rarely showed any side effects. It took a long time to pin down the cause, but it was eventually tracked to a deficiency in an enzyme with the shorthand G6PD, whose gene is carried on the X chromosome. As with sickle-cell anemia, carriers for the G6PD mutations have some resistance to malarial infection, and for this reason G6PD deficiency is more prevalent among people with ancestry from West Africa, where malaria is endemic. This was the first time anyone had noticed that there was a difference in the effects of pharmaceuticals between different racial groups, and it is regarded as the moment when the new field of pharmacogenetics was born. Since then there have been many more examples of severe side effects suffered by different ethnic or racial groups, which has had an influence on drug prescriptions and treatment plans.

Of these the most clinically relevant are the observed differences in the way people metabolize pharmaceuticals. Most drugs, and other toxins, are cleared from the body by the liver, using a series of proteins called P450 cytochromes. Many African Americans carry a version of the P450 gene located on chromosome 10 that is less active in clearing some widely used drugs, like beta-blockers, the blood thinner warfarin, and the anti-inflammatory drug diclofenac. As a consequence, African American patients are generally prescribed much lower doses of these drugs than are their European American counterparts. By now you will begin to see the dangers of the blanket application of this prescribing policy toward anyone classified as African American. If one or both of their P450 genes is actually European in origin, then the basis for prescribing the lower dose will be wrong. A quick scan through the chromosome portraits in the gallery reveals that of my nine African American volunteers, only three have both copies of their P450 gene from African ancestors, three have one European and one African copy, and the genes of the remaining three are completely European. Inversely, "Sugar Kane," my European American volunteer from Florida, also carries an African version of P450 on her chromosome number 10, so even though she is unmistakably white she could well clear drugs more slowly than her physicians would expect using only her overall ethnic affiliation.

But it is not just the collaborations between African and European genes that are highlighted by the portraits. Another of my DNA volunteers, "Holly Golightly," a distinguished African American biographer whom I met while she was on sabbatical in Oxford, was surprised when I told her that both copies of her lactase genes, located on chromosome 2, were inherited from Native American ancestors. She is lactose intolerant and had always put this down to her African background, whereas in fact her inability to break down lactose, which is found mainly in milk, is due to the poorly functioning lactase genes that she inherited from her Native American ancestors.

Possibly the most revealing feature of the chromosome portraits concerns the genes for the one trait that, more than any other has been used to define racial categories—that of color. All pigmentation in humans is due to just one basic substance, melanin. It alone is responsible for the vast range of skin and hair colors found in people from around the world. Melanin itself is a polymer derived from the amino acid tyrosine and is contained within pigmented cells, the melanocytes, in discrete granules. Basically, the more melanocytes and the more melanin in the granules, the darker the skin, eyes, and hair. Blue eyes are not blue because they contain a pigment but because, in the absence of melanin, light reflected from a layer in the iris is diffracted through the regularly spaced transparent collagen fibers in the cornea and gives the appearance of being blue; it is the same optical mechanism that imparts the vivid colors of a butterfly's wing.

The genetic control of skin and hair pigmentation is orchestrated by eleven genes that we know about, though there may well be more. They each control different parts of the process of producing melanin granules and regulating the number of melanocytes. The paler end of the wide range of human pigmentation is probably a response to the reduced exposure to sunlight that some of our ancestors experienced when moving from Africa to higher latitudes. Some vital functions, like the synthesis of vitamin D and folic acid, depend on sunlight, so it makes sense that evolutionary natural selection would have promoted the survival of lighter-skinned individuals. When we look at the chromosome portraits, it is very clear that many of my African American volunteers, who count themselves as black, actually have a mixture of pigmentation genes from many different ancestries. To take just one example, the radio-talk-show host Mark Thompson. Of his eleven pigmentation genes, only two are of completely African origin, five have been inherited equally from European and African ancestors, two are an equal mix of African and Native American, and one has been inherited from exclusively European ancestors. That blend of origins is the direct result of the mixing of his chromosome segments in generations of his African, European, and Native American ancestors.

Since this process is more or less completely random, a vast number of combinations is possible in any African American. There will be individuals who actually have very little DNA from African ancestors, yet if these contributions include chromosome segments housing the pigmentation genes, then they will have typically dark coloring. Likewise there will be Americans whose DNA is almost all African in origin, yet if the pigmentation genes are not included in these segments and instead come from European ancestors, then their coloring will be white. Similarly, it would be entirely possible for a European American with only a small overall component of African DNA to be very dark skinned if these ancestral segments were to include the pigmentation genes. Our only Cherokee volunteer probably had a dark skin tone because the sole surviving segment of Native American DNA in his genome included one of the most influential of the pigmentation genes, located on chromosome 15.

As you leave the gallery, my hope is that you will come away with the feeling that you have glimpsed another world. A world that mocks the artificial divisions we have created for ourselves. A world made up of the corpuscles of DNA that each of us has inherited over millennia from our myriad ancestors, every one of them a resourceful survivor from earlier times. We are their privileged custodians in this world for a few short years, messengers through time to generations not yet born. Let us enjoy this honor while we may.

Acknowledgments

So many people have contributed to *DNA USA*, but let me start by thanking all the volunteers who either allowed me to sample their DNA or shared with me the results of earlier genetic analyses. Without their help there would simply have been nothing to write about. They are, in no particular order, Meriwether Schmid, Christopher Childs, Brenton Simons, Richard Ferguson, Dr. Henry Louis Gates Jr., Dr. Gretchen Holbrook Gertzina, Polly Furbush, Dr. Esteban Burchard, Dr. Roy King, Dr. Rick Kittles, Gina Paige, David Dearborn, Bonnie Healy, Dr. Nanibaa' Garrison, Dr. Justin Barrett, Barbara Poole, Charlie Coleman, Doug Chase, Aaron Gray, Margaretta Barley, Lynda Duncan, Latonya Raston, Justin Connors, Rev. Mark Thompson, Toby Cooper, Brinson Weeks, Sandi Hewlett, Dr. Jay Lewis, and Lee Huntley.

Many people helped by giving up their valuable time to meet or talk with me, including Bennett Greenspan of Family Tree DNA, Dr. Rick Kittles and Gina Paige from African Ancestry, and Dr. Scott Woodward from the Sorenson Institute. I also enjoyed and benefited from conversations with Dr. Kimberly TallBear from the University of California, Berkeley; Dr. Gabrielle Tayac of the National Museum of the American Indian, Washington D.C.; Dr. Mike Hammer from the University of Arizona, Tucson; Anna Silas from the Hopi Cultural Center, Second Mesa, Arizona; Michael Markley, tribal historian of the Seaconke Wampanoag; Juan Luis Castro Suarez, lexicographer and restaurateur; and Serle Chapman, our guide in Wyoming, who provided a wise introduction to the Cheyenne.

I am very grateful for the enthusiastic help with the genome analyses and chromosome painting from 23andMe, Inc. in Mountain View, California—in particular, Dr. Joanna Mountain, Dr. Mike MacPherson, Stewart Ellis, and Linda Avey. Many people made what seemed at first a mammoth task come to completion by their generous welcome and support. I especially want to pay tribute to the New England Historic Genealogy Society, who enthusiastically supported and welcomed me, and my intrepid team of researchers, to their headquarters in Boston. A big thank-you to all the staff at NEHGS and particularly to the president and C.E.O., Brenton Simons; Kelly McCoulf, his tireless PA; and Lynn Betlock, who organized the DNA kits and the recruitment of New England volunteers. Also in Boston, or at least in nearby Cambridge, I owe a great deal to Dr. Henry Louis Gates Jr. and the staff at the W.E.B DuBois Institute for African and African American Studies at Harvard for their most generous hospitality.

All that research and travel had to be turned into a book, and for that I am very grateful to Robin Roberts-Gant and Gerry Black for their invaluable and skillful assistance with the book illustrations. The travel plans were also sometimes complicated, so I owe a lot to Debs Hull of Oxonian Travel, for smoothing the way, and, as so often, to the irreplaceable Hilary Prince and the very talented Mr. Bentley for holding the fort in Oxford while I was in America.

But no book is ever written without encouragement, and for this I must once again thank my literary agent, Luigi Bonomi, and, even more than usual, my editor at W. W. Norton, Bob Weil, whose idea it was in the first place and who, with Philip Marino's editorial help, Sue Llewellyn's and Don Rifkin's copyediting (and translation into American), Chris Carruth's index, Devon Zahn's production skills, and Susan Foden's proofreading, molded my rough and ready manuscript into the finished product.

Lastly, there would simply be no *DNA USA* without Ulla and Richard, who traveled with me over many thousands of miles by road, rail, and air; helped with the sample collection, voice recordings, note taking, and in so many other ways. With Richard's sketches and Ulla's constant encouragement, they made sure the book was finished in style, and on time.

Appendix

Core mutations of Native American mDNA clusters

CLUSTER	MUTATIONS
A1	223 390 319 362
A2	111 223 290 319 362
B	189 217
C	223 298 325 327
D	223 325 362
X	223 278

Calibrated origin dates and clan mother names for Native American mDNA clusters

CLUSTER	MOTHER	AGE (YRS)
A	Aiyana	15,800
B	Ina	18,700
C	Chochmingwu	19,600
D	Djigonese	16,900
X	Xenia	15,800

EUROPEAN

Calibrated origin dates, clan mother names, and frequencies for native European mDNA clusters

CLUSTER	FREQUENCY (%)	MOTHER	AGE (YRS)
U5	5.7	Ursula	47,000
HV	5.4	HV	34,000
X (I)	1.7	Xenia	26,000
U4	3.0	Ulrike	20,000
H	37.7	Helena	14,000
T	2.2	Tara	13,000
K	4.6	Katrine	12,500
T2	2.9	Tara	12,000
J	6.1	Jasmine	8,500
T1	2.2	Tara	9,000

AFRICAN

Clan mothers and ages of native African mDNA clusters

NOTATION	CLAN MOTHER	AGE (YRS)
Superclan L1		
L1A	Layla	40,000
L1B	Lamia	30,000
L1C	Lalamika	60,000
L1D	Latasha	50,000
L1E	Lalla	83,000
L1F	Labana	86,000
L1K	Lakita	92,000
Superclan L2		
L2A	Leisha	55,000
L2B	Lesedi	32,000
L2C	Lingaire	27,000
L2D	Lindewe	122,000
Superclan L3		
L3A	Lara	60,000
L3B	Limber	21,000
L3D	Lingaire	30,000
L3E	Lila	45,000
L3F	Lungile	36,000
L3G	Lubaya	45,000

Regional distribution within Africa of the most frequent native African mDNA clusters

REGION	TOP 5 MOST FREQUENT CLUSTERS
East	L1A, L2, L3A, L3F, L3G
Southeast	L1A, L2A1a, L2A1b, L3E
South	L1A, L1D, L1K, L3B, L3E
Central	L1A, L1C, L2A1, L3E
West	L1B, L2A1, L3B, L3D
North	L1B, L2A1, L3B, L3D

FROM	TO	MUTATIONS
Superclan L1		
Root	L1	16230
Root	L1D	16129, 16243
L1D	L1D1	16294
L1D	L1D2	16234
Root	L1F	16169, 16327
Root	L1A	16129, 16148, 16172, 16188G, 16278, 16320
L1A	L1A1	16168
L1A1	L1A1a	16278
L1A	L1A2	16129
Root	L1K	16172, 16209, 16214, 16291
L1K	L1K1	16166C
Root	L1E	16129, 16148, 16166
L1E	L1E1	16111, 16254, 16311
L1E	L1E2	16355, 16362
L1	L1B/C	7055R
L1	L1B	16126, 16264, 16270
L1B	L1B1	16293
L1B/C	L1C	16129, 16294, 16360
L1C	L1C1	16293
L1C1	L1C1a	16274
L1C1a	L1C1a1	16214, 16223, 16234, 16249
L1C	L1C2	16265C, 16286G
L1C	L1C3	16187, 16215
L1	L2/3	16187, 16189, 16311
Superclan L2		
L2/3	L2	16390
L2	L2C	13957R
L2C	L2C1	16318
L2C	L2C2	16264
L2	L2A	13803R, 16294
L2A	L2A1	16309

L2A1	L2A1a	16286
L2A1	L2A1b	16290
L2	L2D	3693R, 16399
L2D	L2D1	16129, 16189, 16223, 16300, 16354
L2D	L2D2	16111A, 16145, 16239, 16292, 16355
L2	L2B	4157R, 16129, 16114A, 16213
L2B	L2B1	16362

Superclan L3

L2/3	L3	3592R, 16278
L3	L3F	16209, 16311
L3F	L3F1	16292
L3	L3G	16293T, 16311, 16355, 16362
L3	L3B/D	16124
L3B/D	L3B	10084R, 16278, 16362
L3B	L3B1	16124
L3B	L3B2	16311
L3B/D	L3D	8616R
L3D	L3D1	16319
L3D	L3D2	16256
L3D	L3D3	16189, 16278, 16304, 16311
L3	L3E	2349R
L3E	L3E3/4	5260R
L3E3/4	L3E4	16264
L3E3/4	L3E3	16265T
L3E	L3E2	16320
L3E	L3E2b	16172, 16189
L3E	L3E1	16327
L3E1	L3E1a	16185
L3E1	L3E1b	16325
L3A	M	10397R
L3A	N	10871R

Mutations in the African mDNA tree.

Numbers are positions of mutations in the mDNA sequence. Positions without a suffix are transitions. Suffixes A, G, C, T indicate transversions to these bases. Suffix *del* indicates a deletion while *R* is a restriction enzyme variant.

CLAN DONALD GENEALOGY

Identification of individuals on Clan Donald genealogy
(see Fig. 2 on p. 108)

CODE	NAME	DATES	BRANCH
A	Somerled	c. 1100–1164	
B	Dugall, founder of Clan Dougal of Lorne	c. 1118–?	
C	Donald MacRanald of the Isles	1190–1269	
D	Alastair Mor MacDonald, founder of Clan Alastair	d. 1299	
E	John MacDonald, Lord of the Isles	d. 1386	
F	Ranald Macdonald, 1st of Clanranald and Glengarry	d. 1386	
G	Donald Gallach MacDonald, 3rd of Sleat	d. 1506	
H	Ranald Mor MacDonald, 7th of Keppoch	d. 1547	
I	Alan MacDonald, 4th of Clanranald	1437–1481	
J	Donald Gruamach MacDonald, 4th of Sleat	d. 1534	
K	Ian Moidartach MacDonald, 8th of Clanranald	1502–1584	
L	Sir Donald Breac Macdonald, 10th of Sleat	1605–1678	
M	Ranald Alexander MacDonald, 24th of Clanranald	Living	Clanranald
N	P. M. Macdonald	Living	Glenaladale
O	J. J. Macdonald	Living	Bornish
P	Ranald MacDonell, 23rd of Glengarry	Living	Glengarry
Q	Allan Douglas MacDonald of Vallay	Living	Vallay
R	Sir Iain Godfrey MacDonald, 25th of Sleat	Living	Sleat
S	David Foster Macdonald, 17th of Castle Camus	Living	Castle Camus
T	D. B. Macdonald	Living	Sleat
U	W. F. Macdonald	Living	Achnancoichean
V	L. McDonald	Living	Bohuntin

Details of Y chromosome mutations in Clan Donald genealogy (See Fig.2 on p. 108)

LINK	MARKER	MUTATION
C-E	458	16-15
E-G	557	15-16
F-P	449	31-32
F-P	456	17-16
F-P	CDYa	34-35
G-T	GATAH4	12-11
H-U	448	20-21
H-U	449	30-31
H-U	GATAH4	12-11
H-V	570	18-19
H-V	390	25-24
I-O	CDYb	38-37
J-L	448	20-19
J-S	449	31-32
J-S	460	11-10
J-S	557	15-17
K-M	389a	14-13
K-M	390	25-24
K-M	449	31-32
K-M	576	17-18
K-M	CDYb	38-37
K-N	GATAH4	12-10
K-N	CDYa	34-33
K-N	449	31-30
K-N	458	15-16
L-R	607	16-15

Details of individual Y chromosome signatures on Clan Donald genealogy

	1	2	3	4	5	6	7	8	9	10	11	12	13	14	15	16	17	18
C	3	3	1	3	3	3	4	3	4	3	3	3	4	4	4	4	4	4
O	9	9	9	9	8	8	2	8	3	8	9	8	5	5	5	5	5	4
D	3	0		1	5	5	6	8	9	9	2	9	8	9	9	5	4	7
E					a	b				a		b		a	b			

	1	2	3	4	5	6	7	8	9	10	11	12	13	14	15	16	17	18
A	13	25	15	11	11	14	12	12	10	14	11	31	16	08	10	11	11	23
E													15					
F													15					
G													15					
H													15					
I													15					
J													15					
K													15					
L													15					
M		24								13			15					
N													16					
O													15					
P													15					
Q													15					
R													15					
S													15					
T													15					
U													15					
V		24											15					

19	20	21	22	23	24	25	26	27	28	29	30	31	32	33	34	35	36	37	38
437	448	449	464a	464b	464c	464d	460	GATA H4	YCAIIa	YCAIIb	456	607	576	570	CDYa	CDYb	442	438	557
14	20	31	12	15	15	16	11	12	19	21	17	16	17	18	34	38	12	11	15
																			16
																			16
	19																		16
		32											18			37			
		30						10							33				
																37			
		32									16				35				
	19																		16
	19											15							16
		32					10												17
								11											16
21		30						11											
														19					

Code letters in column 1 identify deceased (shaded) and living individuals on Fig. 2
(p. 108). Markers 1–38 are identified in row 2.
Numbers in table body are allele repeat lengths of Y chromosomes that differ from
values deduced for Somerled (row A).

Appendix

The location of 146 important genes, showing their order from right to left along each chromosome (Column 1), the body system involved (Column 2), the encoded protein (Column 3), and the international gene symbol (Column 4).

REF	SYSTEM	PROTEIN	SYMBOL
1.1	Digestion	Chymotrysin C	CTRC
1.2	Bones, skin, teeth, and joints	Bone morphogenetic protein 8a/b	BMP8A/B
1.3	Metabolism	Leptin receptor	LEPR
1.4	Digestion	Amylase	AMY1 and 2
1.5	Digestion	Cathepsin K and S	CTSK/S
1.6	Muscle	Tropomyosin 3	TPM3
1.7	Brain and nerves	Laminin gamma 2	LAMC2
1.8	Liver and kidney	Renin	REN
1.9	Muscle	Actin alpha 1, skeletal muscle	ACTA1
1.10	Eyes	Opsin 3	OPN3
2.1	Immunity	Immunoglobulin kappa constant	IGKC
2.2	Bones, skin, teeth, and joints	Bone morphogenetic protein 10	BMP10
2.3	Digestion	Lactase	LCT
2.4	Brain and nerves	Sodium channel 9	SCN9A
2.5	Heart and lung	Collagen type 3	COL1A3
2.6	Eyes	Crystallin gamma A	CRYGA
2.7	Pigmentation	Fibronectin	FN1
2.8	Eyes	Crystallin beta A2	CRYBA2
2.9	Liver and kidney	Collagen type 4, alpha 3/4	COL4A3/4
2.10	Metabolism	Glucagon	GCG
3.1	Eyes	Retinoic acid receptor beta	RARB
3.2	Metabolism	Parathyroid hormone 1 receptor	PTH1R
3.3	Digestion	Lactotransferrin	LTF
3.4	Metabolism	Phosphodiesterase	PDE12

REF	SYSTEM	PROTEIN	SYMBOL
3.5	Muscle	Myosin HC 15	MYH15
3.6	Eyes	Rhodopsin	RHO
4.1	Brain and nerves	Huntingtin	HTT
4.2	Bones, skin, teeth, and joints	Fibroblast growth factor 3	FGFR3
4.3	Bones, skin, teeth, and joints	Enamalin	ENAM
4.4	Bones, skin, teeth, and joints	Ameloblastin	AMBN
4.5	Immunity	Immunoglobulin J	IGJ
4.6	Liver and kidney	Alcohol dehydrogenase	ADH1A
4.7	Pigmentation	Hair color 2 (red)	HCL2
5.1	Blood	Thrombospondin 4	THBS4
5.2	Muscle	Myosin X	MYOX
5.3	Pigmentation	Solute carrier, family 45, member A2	SLC45A2
5.4	Bones, skin, teeth, and joints	Growth hormone receptor	GHR
5.5	Bones, skin, teeth, and joints	Fibroblast growth factor 10	FGF10
5.6	Bones, skin, teeth, and joints	Fibrillin 2	FBN2
5.7	Bones, skin, teeth, and joints	Lysyl oxidase	LOX
5.8	Bones, skin, teeth, and joints	Fibroblast growth factor 1	FGF1
5.9	Metabolism	Insulin dependent diabetes 10	IDDM10
6.1	Immunity	HLA region	HGC
6.2	Eyes	Opsin 5	OPN5
6.3	Muscle	Myosin VI	MYO6
6.4	Brain and nerves	Laminin alpha 2/4	LAMA2/4
6.5	Brain and nerves	Opioid receptor mu 1	OPRM1
6.6	Blood	Plasminogen	PLG
6.7	Blood	Thrombospondin 2	THBS2
7.1	Muscle	Actin beta	ACTB

REF	SYSTEM	PROTEIN	SYMBOL
7.2	Muscle	Myosin light chain 7	MYL7
7.3	Bones, skin, teeth, and joints	Elastin	ELN
7.4	Blood	Erythropoietin	EPO
7.5	Bones, skin, teeth, and joints	Collagen type 1, alpha 2	COL1A2
7.6	Brain and nerves	Acetylcholinesterase	ACHE
7.7	Eyes	Opsin 1 cone short wave	OPN1SW
7.8	Brain and nerves	Olfactory receptor 2A.25	OR2A25
8.1	Heart and lung	Surfactant protein C	SFTPC
8.2	Bones, skin, teeth, and joints	Fibroblast growth factor receptor 1	FGFR1
8.3	Blood	Plasminogen activator (tissue)	HBB
8.4	Brain and nerves	Opioid receptor kappa 1	OPRK1
8.5	Blood	Angiopoietin 1	ANGPT1
8.6	Brain and nerves	Otoconin	OC90
9.1	Immunity	Inteferon alpha	IFNA
9.2	Muscle	Tropomyosin 2	TPM2
9.3	Digestion	Beta glucosidase (bile acid)	GBA2
9.4	Bones, skin, teeth, and joints	Osteoclast stimulating factor 1	OSTF1
9.5	Bones, skin, teeth, and joints	Osteoglycin	OGN
9.6	Blood	Hemogen	HEMGN
9.7	Blood	ABO blood group	ABO
10.1	Eyes	Optineurin	OPTN
10.2	Muscle	Myosin IIIA	MYO3A
10.3	Eyes	Opsin 4	OPN4
10.4	Muscle	Actin alpha 2, smooth muscle	ACTA2
10.5	Digestion	Trypsin	TYSND1
10.6	Heart and lung	Adrenergic alpha 2A/ beta receptor	ADRA2A/B1

REF	SYSTEM	PROTEIN	SYMBOL
10.7	Metabolism	Cytochrome P450, family 2C, polypeptide	CYP2C9
10.8	Heart and lung	Pancreatic lipase	PNLIP
11.1	Blood	Beta globin	HBB
11.2	Metabolism	Insulin	INS
11.3	Brain and nerves	Otogelin	OTOG
11.4	Digestion	Pepsin	PGA3
11.5	Digestion	Cathepsin C	CTSC
11.6	Pigmentation	Tyrosinase	TYR
11.7	Eyes	Crystallin alpha B	CRYAB
11.8	Immunity	Thymocyte cell surface antigen	THY1
12.1	Immunity	Alpha 2 macroglobulin	A2M
12.2	Digestion	Islet amyloid polypeptide	IAPP
12.3	Pigmentation	Keratin 1-5	KRT1-5
12.4	Bones, skin, teeth, and joints	Collagen type 2	COL2A1
12.5	Digestion	Peptidase B	PEPB
12.6	Pigmentation	KIT ligand	KITLG
12.7	Immunity	Thympoietin	TMPO
12.8	Digestion	Phopholipase A2	PLA2G1B
13.1	Brain and nerves	Olfactomedin 4	OLFM4
13.2	Brain and nerves	Olfactory receptor E	OR7E
13.3	Muscle	Myosin XVI	MYO16
13.4	Liver and kidney	Collagen type 4, alpha 1 and 2	COL4A1/2
14.1	Heart and lung	Myosin heavy chain 6 and 7, cardiac	MYH6/7
14.2	Metabolism	Estrogen receptor 2	ESR2
14.3	Bones, skin, teeth, and joints	Bone morphogenetic protein 4	BMP4
14.4	Blood	Spectrin beta	SPTB
14.5	Digestion	Chymotrypsin	CTRB1
14.6	Brain and nerves	Creatine kinase (brain)	CKB

REF	SYSTEM	PROTEIN	SYMBOL
14.7	Immunity	Immunoglobulin H alpha 1	IGHA1 et al.
15.1	Muscle	Actin alpha 1, cardiac muscle	ACTC1
15.2	Pigmentation	Oculocutaneous albinism II	OCA2
15.3	Brain and nerves	Cholinergic receptor nicotinic alpha 7	CHRNA7
15.4	Pigmentation	Solute carrier, family 24 member 5	SLC24A5
15.5	Blood	Thrombospondin 1	THBS1
15.6	Muscle	Myosin 1E	MYO1E
15.7	Pigmentation	Hair color 3 (brown)	HCL3
15.8	Bones, skin, teeth, and joints	Chondroitin sulfate proteoglycan 4	CSPG4
15.9	Bones, skin, teeth, and joints	Aggrecan	ACAN
16.1	Blood	Alpha globin	HBA1 and 2
16.2	Digestion	Myosin HC 11, smooth muscle	MYH11
16.3	Blood	Alpha hemoglobin stabilizing protein	AHSP
16.4	Muscle	Myosin light chain kinase 3	MYLK3
16.5	Blood	Haptoglobin	HP
16.6	Digestion	Chymotrypsinogen B1/2	CTRB1/2
16.7	Pigmentation	Melanocortin 2 receptor	MC2R
17.1	Muscle	Myosin, heavy chain 1, skeletal adult	MYH1-4
17.2	Eyes	Crystallin beta A1	CRYBA1
17.3	Bones, skin, teeth, and joints	Collagen type 1, alpha 1	COL1A1
17.4	Digestion	Gastrin	GAST
17.5	Heart and lung	Angiotensin 1 converting enzyme	ACE
17.6	Metabolism	Glucagon receptor	GCGR
17.7	Muscle	Actin gamma	ACTG1

REF	SYSTEM	PROTEIN	SYMBOL
18.1	Pigmentation	Melanocortin 2 receptor	MC2R
18.2	Metabolism	Insulin-dependent diabetes 6	IDDM6
18.3	Digestion	Gastrin-releasing peptide	GRP
18.4	Brain and nerves	Myelin basic protein	MBP
19.1	Metabolism	Insulin receptor	INSR
19.2	Pigmentation	Hair color (1 brown)	HCL1
19.3	Brain and nerves	Myelin-associated glycoprotein	MAG
19.4	Digestion	Gastric inhibitory poly-peptide receptor	GIPR
20.1	Liver and kidney	Arginine vasopressin	AVP
20.2	Pigmentation	Agouti signaling peptide	ASIP
20.3	Metabolism	Topoisomerase 1	TOP1
20.4	Bones, skin, teeth, and joints	Bone morphogenetic protein 7	BMP7
20.5	Bones, skin, teeth, and joints	Collagen, type IX, alpha 3	COL9A3
21.1	Bones, skin, teeth, and joints	Chondrolectin	CHODL
21.2	Eyes	Crystallin alpha A	CRYAA
22.1	Immunity	Immunoglobulin lambda constant	IGKC
22.2	Eyes	Crystallin beta A4	CRYBA4
22.3	Brain and nerves	Synaptogyrin 1	SYNGR1

Notes

Chapter 1: The Point of Clovis

1. R. Gramly et al., *Archaeology of Eastern North America* 18 (1988), 5.
2. J. Adovasio and R. Carlisle, *Science* 239 (1988), 713.
3. T. Dillehay and M.B. Collins, *Nature* 332 (1988), 150.
4. T. Dillehay et al., *Science* 320 (2008), 784.
5. J. Adovasio and J. Page, *The First Americans* (New York: Modern Library, 2003), 209.
6. M. Gilbert et al., *Science* 320 (2008), 786.
7. M. Waters and T. W. Stafford Jr., *Science* 315 (2007), 1122.
8. C. Haynes et al., *Science* 317 (2007), 320.

Chapter 2: The Nature of the Evidence

1. S. Paabo, *Nature* 314 (1985), 644.
2. R. Higuchi et al., *Nature* 312 (1984), 282.
3. E. Hagelberg, R. Hedges, and B. Sykes, *Nature* 342 (1989), 485.
4. S. Paabo, in *The Human Inheritance: Genes, Language, and Evolution*, ed. B. Sykes (Oxford: Oxford University Press, 1999), 119.
5. R. Green et al., *Science* 328 (2010), 710.
6. M. Rasmussen et al., *Nature* 463 (2010), 757.

7.   J. S. Jones, *Nature* 314 (1985), 576.

8.   G. Church, quoted in E. Pennisi, *Science* 328 (2010), 683.

Chapter 3: The First Americans

1.   R. Ward et al., *Proceedings of the National Academy of Science (USA)* 88 (1991), 8720.

2.   Principally Martin Richards and Vincent Macaulay.

3.   P. Forster et al., *American Journal of Human Genetics* 59 (1996), 935.

4.   U. Perego et al., *Current Biology* 10 (2009), 1.

5.   M. Hurles et al., *American Journal of Human Genetics* 72 (2003), 1282.

6.   R. Cann, *American Journal of Human Genetics* 55 (1994), 7.

7.   P. Orr, *Science* 135 (1962), 219.

8.   T. Dillehay et al., *Science* 320 (2008), 784.

Chapter 4: The Mystery of Cluster X

1.   R. Scozzari et al., *American Journal of Human Genetics* 60 (1997), 241.

2.   M. Brown et al., *American Journal of Human Genetics* 63 (1998), 1852.

3.   A. Stone and M. Stoneking, *American Journal of Human Genetics* 62 (1998), 1153.

4.   W. Hauswirth et al., *Experientia* 50 (1994), 585.

5.   B. Sykes et al., *American Journal of Human Genetics* 57 (1995), 1463.

6.   R. Taylor et al., *Science* 280 (1998), 1171.

7.   F. Kaestle et al., www.nps.gov/archeology/kennewick/kaestle.htm, 2000.

8.   R. Dalton, *Nature* 420 (2002), 111.

9.   J. Neel, *American Journal of Human Genetics* 14 (1962), 353.

10. R. McInnes, *American Journal of Human Genetics* 88 (2012), 254.

Chapter 5: The Europeans

1.   M. Richards et al., *American Journal of Human Genetics* 67 (2000), 1251.

Chapter 6: The Genetic Genealogy Revolution

1.   E. Foster et al., *Nature* 396 (1998), 27.

2.   B. Sykes and C. Irven, *American Journal of Human Genetics* 66 (2000), 1417.

3.   M. Thomas et al., *Nature* 394 (1998), 139.

Chapter 7: The World's Biggest Surname Project

1. http://dna-project.clan-donald-usa.org.
2. L. Moore et al., *American Journal of Human Genetics* 78 (2006), 334.

Chapter 8: The Jews

1. *The Jewish Encyclopedia* (New York: Funk & Wagnalls, 1901–6).
2. S. Reuter, *Canadian Journal of Sociology* 31 (2006), 291.
3. S. Adams et al., *American Journal of Human Genetics* 83 (2008), 725.

Chapter 9: The Africans

1. Quoted in G. H. Gerzina, *Black London* (New Brunswick, NJ: Rutgers University Press, 1995), 3.
2. E. Watson et al., *American Journal of Human Genetics* 61 (1997), 691.
3. R. Cann et al., *Nature*, 325 (1987), 31.

Chapter 11: All My Ancestors

1. D. Rohde et al., *Nature* (2004), 562.
2. J. S. Jones, *Daily Telegraph (London)*, April 20, 2009.
3. www.genomesonline.org/cgi-bin/GOLD.
4. International HapMap Consortium, *Nature* 437 (2005), 1299; International Hap-Map Consortium, *Nature* 449 (2007), 851.

Chapter 16: The Call of the Canyon

1. A. Silas, *Journey to Hopi Land* (Tucson, AZ: Rio Nuevo, 2006).

Chapter 17: A Question of Blood

1. J. Carter, "American Indian Religious Statement on Signing S.J. Res. 102 into Law," August 12, 1978.
2. D. Bolnick et al. *Science* 399 (2007), 318.
3. K. TallBear, *Journal of Law, Medicine, and Ethics* 35 (2007), 412.
4. A. Harmon, "Tribal Enrollment Councils," *Western Historical Quarterly* 32 (2001), 175.

5.   K. TallBear, *Wicazo Sa Review* 18 (2003), 81.

6.   F. Flam, *San Diego Union-Tribune*, October 5, 2005 (www.signonsandiego.com/uniontrib/20051005/news_lzlc05schurr.html).

7.   S. Zhadanov et al., *American Journal of Physical Anthropology* 142 (2010), 579.

8.   B. Koerner, *Wired*, September 13, 2005 (www.wired.com/wired/archive/13.09/seminoles.html).

Chapter 18: Portraits of America

1.   T. Cooper, *White Women* (Ashburn, VA: Cooper Corporate Publishing, 1998).

Chapter 19: The Private View

1.   A. E. Mourant, *The Distribution of Human Blood Groups* (Oxford: Blackwell Scientific Publications, 1954).

Index

Page numbers in *italics* refer to illustrations.

About the Author

Bryan Sykes, a professor of human genetics at the University of Oxford and a Fellow at Wolfson College, has had a remarkable scientific career. He was the first to discover, in 1989, how to recover DNA from human remains thousands of years old, and he has been called in as the leading international authority to examine several high-profile cases, such as the Ice Man, Cheddar Man, and the many individuals claiming to be surviving members of the Russian royal family. Since then he has worked extensively on the origins of peoples from all over the world, using DNA from living people as well as from archaeological remains. He has proved that the origin of Polynesians lay in Asia, not America, and discovered that the ancestors of most Europeans were hunter-gatherers from before the last Ice Age. He also has showed that most Europeans trace their maternal genetic ancestry back to only seven women. On the male side, he was the first to show the close connection between DNA and surnames, a discovery that is revolutionizing genealogy.

Sykes is the founder and chairman of Oxford Ancestors (www.oxford ancestors.com). the first company in the world to help people explore their own genetic roots using DNA. He is also the author of *The Seven Daughters of Eve, Adam's Curse,* and *The Human Inheritance: Genes, Language, and Evolution.*